Graph Theory and Its Applications to Problems of Society

FRED S. ROBERTS
Rutgers University

SOCIETY for INDUSTRIAL and APPLIED MATHEMATICS • 1978

PHILADELPHIA, PENNSYLVANIA 19103

Library of Congress Cataloging in Publication Data

Roberts, Fred S

Graph theory and its applications to problems of society.

(Regional conference series in applied mathematics; 29)

"Based on a series of ten lectures delivered at a regional conference . . . held at Colby College on June 20-24, 1977."

Includes bibliographical references and index.

1. Social sciences–Mathematical models–Congresses. 2. Graph theory–Congresses. 3. Social problems–Mathematical models–Congresses. I. Title. II. Series.

H61.R59 300'.1'51 78-6277

Printed for the Society for Industrial and Applied Mathematics by

J. W. Arrowsmith Ltd., Bristol 3, England

Contents

Preface

This monograph is based on a series of ten lectures delivered at a regional conference on Graph Theory and its Applications to Problems of Society held at Colby College on June 20–24, 1977. The conference was sponsored by the Conference Board of the Mathematical Sciences and the National Science Foundation.

I wish to thank everyone at Colby College for their hospitality, and especially Professor Lucille Zukowski who planned and organized the conference and persevered until her goals became a reality. My special thanks go to Lynn Braun, who nobly typed the notes for the conference on short notice, and to Robert Opsut, who proofread them. I also wish to thank the National Science Foundation for support of the research on which parts of these lectures were based.

Finally, I wish to thank my wife Helen, not only for her insightful professional assistance, but for her unselfish personal encouragement, and my new-born daughter Sarah, who is probably the only one who learned more during the time these lectures were prepared than I did.

FRED S. ROBERTS

CHAPTER 1
Introduction

1.1. The scope of the work. Sometimes it seems that our society faces overwhelmingly difficult problems, problems involving energy, transportation, pollution, perturbed ecosystems, urban services, the economy, genetic changes, social inequalities, and so on. Increasingly, mathematics is being used, at least in small ways, to tackle these problems. In these lectures we shall examine the role of one branch of mathematics, graph theory, in applications to such problems of society.

We have chosen to present mathematical topics from the field of graph theory because graphs have wide-ranging applicability and because it is possible in graph theory to bring a previously unfamiliar scientist to the frontiers of research rather quickly. The choice of topics from within graph theory and even more so the order of presentation of these topics is not typical of the graph theory literature. Rather, the topics were chosen to best illustrate the applications, and to lead into them as quickly as possible. Some of the more traditional topics of graph theory, such as colorability, independence, and eulerian chains, are not covered until fairly late. Then, they are presented with an emphasis on results of applied interest.

We have tried to be self-contained in preparing these notes. However, they are written on a research level, with the goal being to present results at the frontiers of current graph-theoretical work. The reader will find some of these same topics discussed at a more leisurely pace in Roberts (1976a). These lectures are in some sense a continuation of the topics presented in Chapters 3 and 4 of that book. They go beyond the results stated there, present more recent work, and introduce a variety of additional applied and graph-theoretical topics.

The problems of society with which we are concerned are extremely complex and wide-ranging. At the outset, let us put into perspective the role of graph theory in particular and mathematics in general vis-à-vis these problems. We will not claim that graph theory alone can solve these problems. Nor will we argue that they cannot be solved without graph theory. Rather, we hope to demonstrate that the use of precise, graph-theoretical reasoning can cast light on such problems, provide tools to help in making decisions about them, and help in finding answers to a variety of specific questions which arise in the attempt to tackle the broader issues.

Graph theory is a tool for formulating problems, making them precise, and defining fundamental interrelationships. Sometimes, as we shall see, simply formulating a problem precisely helps us to understand it better. The very act of formulation is an aid to understanding. In this way, graph theory plays the role of a learning device, which we can use as an aid in pinpointing future directions and approaches. We shall see this role for graph theory, for example, in our discussion of energy modeling. Once a problem has been formulated in graph-theoretical language, the concepts of graph theory can be used to define concepts which are useful in analyzing the problem. Also, as we shall see for example in our discussion of balance theory and social inequalities, graph theory can lead to new theoretical concepts which can be used to build theories about social problems. Often, formulation of the problem precisely is enough to give us insight on why the problem is hard. For example, formulation of a problem posed by the New York City Department of Sanitation as a graph coloring problem suggests why the problem is difficult: coloring of graphs is a hard problem in a precise sense.

Of course, graph theory has uses beyond simple problem formulation. Sometimes a part of a large problem corresponds exactly to a graph-theoretic problem, and that problem can be completely solved. We shall see this with a certain telecommunications problem, for example. We shall also see it with problems of seriation and measurement. Here, solution to the graph-theoretic problem posed provides tools for organization of data in archaeology, psychology, and in decision and policy problems. Sometimes, once a problem is formulated in graph-theoretical language, we will discover that the problem is hard to solve. We will discover this, for example, when we discuss how to orient streets so as to move traffic efficiently, and when we discuss the dimensionality of ecological phase space using graph theory. In both cases, even when enough simplifying assumptions are made to state a problem graph-theoretically, that problem has not been solved and is at the frontiers of current mathematical research. Thus, surprisingly, sometimes it is the lack of mathematical knowledge which is a limiting factor.

Usually if a problem is formulated graph-theoretically, it is done so as the result of simplifications, for example, the omission of important aspects such as changing relationships over time or strengths of effects. Sometimes these simplifications are not significant. We shall see this, for example, with problems of traffic light phasing, street sweeping, committee scheduling, and the assignment of mobile radio frequencies. In all of these cases, the solutions to the corresponding graph-theoretical problem are rather quickly amenable to practical application. In other cases, because of the significance of the simplifying assumptions, the conclusions from solution of a graph-theoretic problem can only be taken as tentative and suggestive. However, these solutions can suggest useful strategies to consider and future directions in which to investigate. Moreover, graph-theoretical analysis can help to pinpoint the simplifying assumptions and suggest promising directions which can remove these.

If all of this seems to suggest that graph theory is a panacea and by itself can solve a large number of problems, let us quickly disclaim that suggestion. Graph theory is just one tool, which sometimes solves problems and sometimes gives us insights. It usually has to be used along with many other tools, mathematical and otherwise. Hopefully, the use of graph theory can help us to understand in small ways the large problems which face our society, and some of their possible solutions.

Finally, let us remember that applied mathematics develops in close contact with applications. Many of the results stated in the following as purely mathematical results were motivated by specific applied questions. Problems of society have been a stimulus to the development of new mathematics, and should continue to be one in the future.

1.2. Digraphs and graphs. We shall adopt the terminology and notation of Roberts (1976a). In particular, a *directed graph* or *digraph* will consist of a finite set V of *vertices* and a set A of ordered pairs of vertices called *arcs.* We shall represent a digraph (V, A) by drawing the vertices as points and drawing an arrow from x to y if and only if the ordered pair (x, y) is an arc. We shall usually not allow *loops,* arcs of the form (x, x). However, there are several places where loops will be a convenience, and we shall explicitly allow them there. If a digraph is *symmetric* in the sense that $(x, y) \in A$ if and only if $(y, x) \in A$, then we shall usually replace the two arrows between x and y with an undirected line. We may think of a symmetric digraph as a set V of vertices together with a set E of unordered pairs of vertices, the elements of E being called *edges.* The pair (V, E) will be called a *graph.*

Let us give some examples.

Example 1. Suppose we let the vertices be locations in a city and draw an arc from x to y if there is a street leading from x to y. Then we obtain a digraph, which may or may not be symmetric. We shall be interested in various rearrangements of directions on streets which might allow traffic to move more efficiently and hence cut down on air pollution.

Example 2. In Example 1, let certain of the arcs be designated as streets to be cleaned during a given time period. We shall be interested in finding a route for a street cleaner which cleans all of the designated streets in the shortest amount of time.

Example 3. Let the vertices be species in an ecosystem. Draw an arc from x to y if species x preys on species y. The resulting *food webs* are digraphs. Similarly, we can obtain a graph from the species by joining two species with an edge if and only if they compete for a common prey. We shall be interested in what we can learn about the relation between the food web and the competition graph; the results will tell us something about the number of factors required to understand competition or niche overlap in ecosystems.

Example 4. Let the vertices of a digraph be locations in a nuclear power plant. Draw an arc from location x to location y if it is possible for a watchman at x to see a warning light at y. We shall be interested in finding a minimal number of watchmen who can oversee all the locations.

Example 5. Let various traffic streams or directions of traffic be vertices of a graph and draw an edge from x to y if the two traffic streams x and y are compatible. We shall use this compatibility graph to phase traffic lights.

Example 6. Let the vertices be factors relevant to energy demand, and draw an arc from x to y if a change in x has a significant effect on y. By considering the direction of the effect (increasing or decreasing), we shall try to make qualitative forecasts of future levels of variables such as energy demand.

Example 7. Let the vertices of a graph be alternatives among which an individual is expressing his preferences, and draw an edge between two vertices if and only if the individual is indifferent between the two alternatives. We shall use this indifference graph to measure the individual's opinions.

Example 8. Let the vertices of a graph be possible codewords for a rapid communication system, and draw an edge between two codewords if it is possible to confuse them. This confusion graph will be used to find codes with large capacity.

Example 9. Let the vertices of a digraph be certain "extended bases" from an RNA chain. Draw an arc from extended base x to extended base y if in a complete digest of the chain, there is a fragment of the chain beginning in x and ending in y. The resulting digraph will be used in recovering information about the structure of the chain.

1.3. Reaching and joining. Let $D=(V, A)$ be a digraph. We shall be concerned with ways to reach one vertex from another. We shall adopt the following terminology about reachability. A *path* in D is a sequence $u_1, a_1, u_2, a_2, \cdots, u_t, a_t, u_{t+1}$, where each u_i is a vertex, each a_i is an arc, and a_i is the arc (u_i, u_{i+1}). The path has *length t*. For example, in the digraph D_1 of Fig. 1.1, a, (a, b), b, (b, c), c, (c, e), e, (e, b), b is a path of length 4, which is unambiguously abbreviated as a, b, c, e, b. A path is *simple* if it has no repeated vertices. It is *closed* if $u_{t+1}=u_1$. It is a *cycle* if it is closed and $u_1, u_2, \cdots, u_t$ are

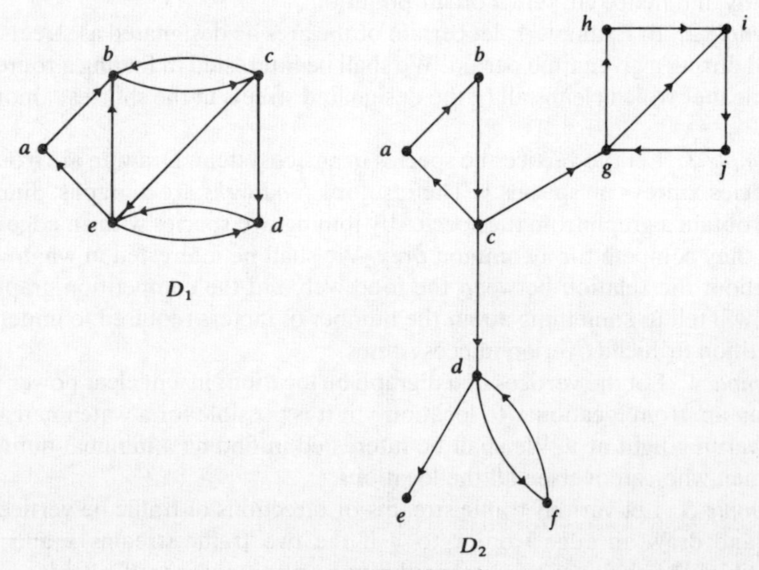

FIG. 1.1. *Two digraphs.*

distinct. In digraph D_1 of Fig. 1.1, *a, b, c, e, b* is not a simple path, while *e, a, b, c,* is; *c, b, c, e, b, c* is a closed path which is not a cycle; and *b, c, d, e, b* is a cycle (of length 4).

There are analogous notions for a graph $G=(V, E)$. A *chain* in G is a sequence $u_1, e_1, u_2, e_2, \cdots, u_t, e_t, u_{t+1}$, where each u_i is a vertex, each e_i is an edge, and e_i is the edge $\{u_i, u_{i+1}\}$. The *length* of the chain is t. A chain is *simple* if it has no repeated vertices. It is *closed* if $u_{t+1}=u_1$. Finally, it is a *circuit* if it is closed, if $u_1, u_2, \cdots, u_t$ are distinct, and if $e_1, e_2, \cdots, e_t$ are distinct.[1] In the graph G_1 of Fig. 1.2, $a, \{a, b\}, b, \{b, e\}, e, \{e, c\}, c, \{c, b\}, b, \{b, e\}, e$ is a chain of length 5, which can be unambiguously written as *a, b, e, c, b, e*. This chain is not simple. An example of a closed chain which is not a circuit is *b, c, e, d, c, e, b* and an example of a circuit is *b, c, d, e, b*. Notice that *a, b, a* is not a circuit, even though $u_1, u_2, \cdots, u_t$ are distinct.

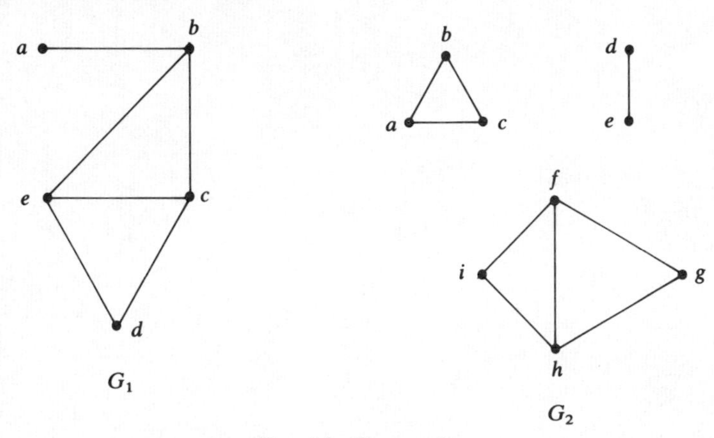

FIG. 1.2. *Two graphs.*

1.4. Connectedness. We shall say a digraph is *strongly connected* if for every pair of vertices x and y, there is a path from x to y and a path from y to x. We say a graph is *connected* if between every pair of vertices there is a chain. In Fig. 1.1, digraph D_1 is strongly connected, while digraph D_2 is not, since, for example, there is no path from d to c. In Fig. 1.2, graph G_1 is connected and graph G_2 is not, since, for example, there is no chain from a to d.

The relation defined by xSy if x is reachable from y by a path and y is reachable from x by a path is an equivalence relation. Hence, the vertices of a digraph are split under this relation into equivalence classes, called *strong components.* Similariy, in a graph, the equivalence classes under the relation "are joined by a chain" form what are called *components* or *connected components.* In digraph D_2 of Fig. 1.1, $\{a, b, c\}$, $\{d, f\}$, $\{e\}$, and $\{g, h, i, j\}$ form the strong components. In graph G_2 of Fig. 1.2, $\{a, b, c\}$, $\{d, e\}$, and $\{f, g, h, i\}$ form the components.

[1] In a digraph, distinctness of the arcs $a_1, a_2, \cdots, a_t$ in a cycle follows from distinctness of the vertices $u_1, u_2, \cdots, u_t$.

CHAPTER 2
The One-Way Street Problem

2.1. Robbins' theorem. The first problem we consider has to do with movement of traffic. If traffic were to move more rapidly and with fewer delays in our cities, this would alleviate wasted energy and air pollution (from idling or stop and go driving). It has sometimes been argued that making certain streets one-way would move traffic more efficiently. In this section, we consider the problem of whether or not it is possible to make certain designated streets one-way and, if so, how to do it.

Of course, it is always possible to make certain streets in a city one-way. Simply put up a one-way street sign! What is desired is to do so in such a way that it is still possible to get from any place to any other place. Let us begin with the simplified problem where every street is currently two-way and it is desired to make every street one-way in the future. We can formulate this problem graph-theoretically by taking the street corners as the vertices of a graph, and drawing an edge between two street corners if and only if these corners are currently joined by a two-way street. We wish to place a direction on each edge of this graph—we speak of *orienting* each edge—so that in the resulting digraph, it is possible to go from any place to any other place, i.e., so that the resulting digraph is strongly connected.

Does every graph G have a strongly connected orientation? Of course, G must be connected to start with—there must be a chain from any vertex to any other. But is that the only condition required? The answer is no. The graphs of Fig. 2.1 are all connected, but none have a strongly connected orientation. For,

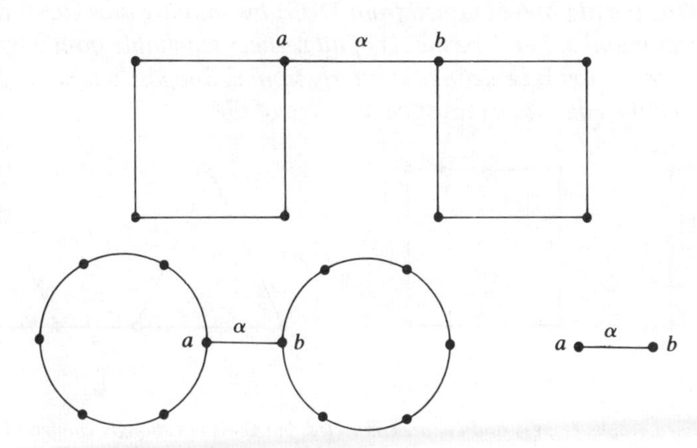

FIG. 2.1. *Connected graphs with no strongly connected orientation.*

in each case, there is a problem with an orientation of the edge labeled α. If this edge is oriented from a to b, then there is no path from b to a; and if the edge is oriented from b to a, then there is no path from a to b.

We say an edge α in a connected graph G is a *bridge* if removal of α, but not its end vertices, results in a disconnected graph. All of the edges α in the graphs of Fig. 2.1 are bridges. It is easy to see that if a connected graph has a strongly connected orientation, then it cannot have any bridges. The converse of this result is true as well, and this will be proved in the next section.

THEOREM 2.1 (Robbins (1939)). *A graph G has a strongly connected orientation if and only if G is connected and has no bridges.*

2.2. Some streets two-way. Let us now consider the case where some streets are to be made one-way, while others remain two-way. Let D be a digraph with vertex set a set of street corners and an arc from x to y if there is a street joining x and y and it is permissible to drive from x to y. It will be convenient to replace the two arcs on a two-way street with one undirected edge. The resulting object, consisting of a set of vertices, some joined by one-way arcs and some joined by undirected edges, will be called a *mixed graph G.* The digraph D will be called the *digraph underlying G*, and will be denoted $D(G)$. We say a mixed graph is *strongly connected* if $D(G)$ is strongly connected. We say a mixed graph G is *connected* if, when we disregard direction on arcs, we obtain a connected graph. An undirected edge α in a mixed graph is a *bridge* if removal of α but not its end vertices results in a mixed graph which is not connected. For example, in Fig. 2.2, mixed graphs G_1, G_2 and G_3 are all connected, but G_2 is not strongly connected. Edge α in G_3 is a bridge.

THEOREM 2.2 (Boesch and Tindell (1977)). *Suppose G is a strongly connected mixed graph. Then for every edge $\{u, v\}$ of G which is not a bridge, there is an orientation of $\{u, v\}$ so that the resulting mixed graph is still strongly connected.*

The proof depends on a lemma.

LEMMA. *Suppose G is a strongly connected mixed graph and $\{u, v\}$ is an edge of G. Let D′ be the digraph obtained from D(G) by omitting arcs (u, v) and (v, u) but not vertices u and v. Let A be the set of all vertices reachable from u by a path in D′, less the vertex u. Let B be defined similarly from v. Suppose u is not in B and v is not in A. Then the edge $\{u, v\}$ must be a bridge of G.*

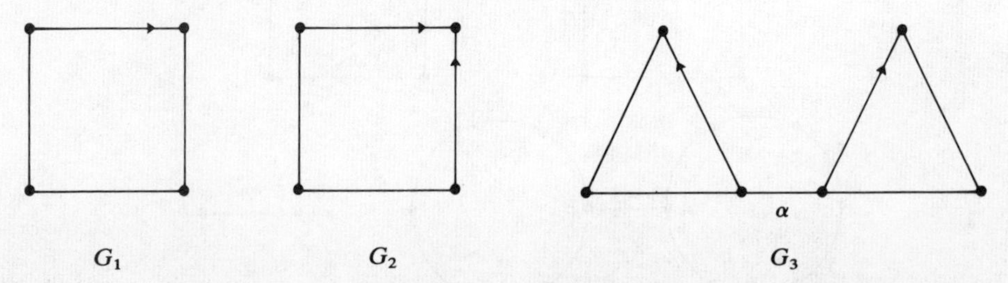

FIG. 2.2. *Mixed graphs G_1, G_2 and G_3 are connected, but G_2 is not strongly connected. Edge α in G_3 is a bridge.*

Proof. Let us observe first of all that every vertex of G is in A or in B. For if w is not in A or in B, then in $D(G)$ there could be no paths from u to w or v to w, and hence G is not strongly connected. Next, A and B must be disjoint. For, suppose we are given w in $A \cap B$. By definition of A and B, $w \neq u, v$. By strong connectedness, there is in $D(G)$ a path from w to u and a path from w to v. It follows that there is a path from w to u in D' or there is a path from w to v in D'. In the former case, there is a path from v to w to u in D'; hence u is in B. In the latter case, v is in A. Finally, we observe that in G less the edge $\{u, v\}$, there could be no arc or edge joining vertices of A and of B. For, if there were such an arc or edge, there would be an arc in D' of the form (u', v') or (v', u'), for u' in A and v' in B. In the former case, v' would have to be in A, contradicting the fact that A and B are disjoint. In the latter case, we get a similar contradiction. Now, we have shown that A and B partition the vertices of G and in G less the edge $\{u, v\}$ there is no arc or edge joining vertices of these two sets. Hence, G less $\{u, v\}$ is not connected, and so $\{u, v\}$ is a bridge. Q.E.D.

Having proved this lemma, we now prove Theorem 2.2. By the lemma, either u is in B or v is in A. If the former, then orienting $\{u, v\}$ from u to v results in strong connectedness, and in the latter, orienting $\{u, v\}$ from v to u results in strong connectedness. If both, either orientation works. This proves the theorem. A proof of the missing direction of Theorem 2.1 follows by orienting one edge at a time, and noting that at each stage, if $\{u, v\}$ is not a bridge, it could not become one after the orientation.

2.3. Algorithms for one-way street assignments. Theorem 2.1 is not a very practical result unless it is accompanied by an algorithm for obtaining a one-way street assignment. Our proof of Theorem 2.1, however, suggests that we may carry out the orientation one step at a time. This procedure is carried out in Fig. 2.3. Note that the first two choices of orientation were arbitrary. However, the third choice was forced, as once these two orientations are chosen as shown, there is no path in the mixed graph from b to a once the edge between a and b is deleted. Similarly, all subsequent choices of orientation are forced.

An alternative algorithm for obtaining a strongly connected orientation is based on the method of *depth first search.* The procedure is to label vertices with the integers $1, 2, \cdots, n$, where n is the number of vertices of the original connected, bridgeless graph. Start by picking a vertex at random and labeling it 1. Pick any (unlabeled) vertex joined to the vertex labeled 1 by an edge, and label it 2. In general, having labeled vertices with the labels $1, 2, \cdots, k$, search through all vertices one step away from that vertex labeled k. If there is such a vertex which is unlabeled, pick one and label it $k+1$. Otherwise, find the highest label j so that there is an unlabeled vertex one step from j, pick such a vertex, and label it $k+1$. This labeling procedure is carried out in Fig. 2.4. Note that after the label 4 has been used, we must return to the vertex labeled 2 before being able to find an unlabeled vertex. Note that the labeling procedure can be completed if and only if one starts with a connected graph. After the labeling has been carried out, a strongly connected orientation is obtained by orienting an

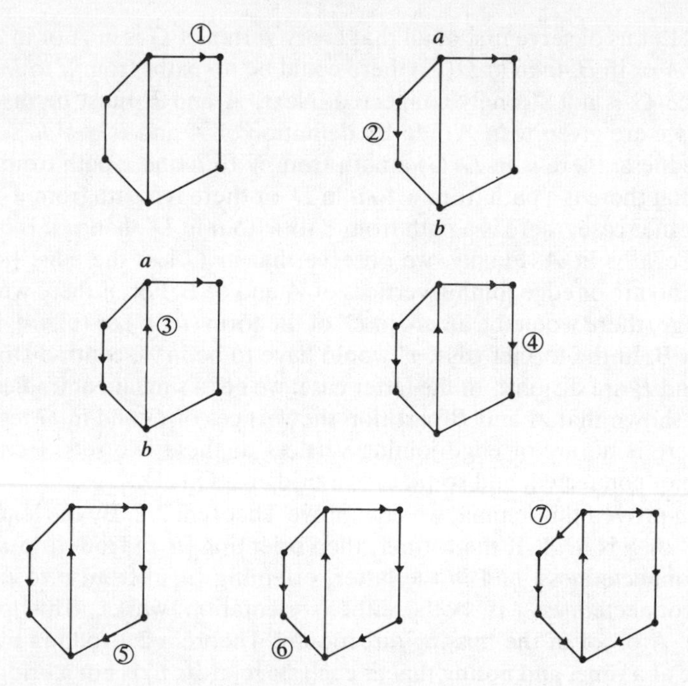

FIG. 2.3. *The construction of a strongly connected orientation using the Boesch–Tindell procedure. Steps ① and ② are arbitrary and the remaining steps are forced.*

edge from lower number to higher number if it was used in the labeling procedure, and otherwise orienting from higher number to lower number. This orientation corresponding to the labeling shown in Fig. 2.4 is illustrated in Fig. 2.5. See Roberts (1976a) for a proof that this orientation will always be strongly connected if one starts with a connected, bridgeless graph.

The second algorithm appears to be faster than the first. In depth first search, we have $|V(G)|$ steps, one corresponding to each assignment of label. At each step, we investigate a certain number of edges. Since we can be sure that no edge previously investigated is again investigated, the total number of investigations of edges is $|E(G)|$. Thus, the labeling procedure takes on the order of $|V(G)|+$

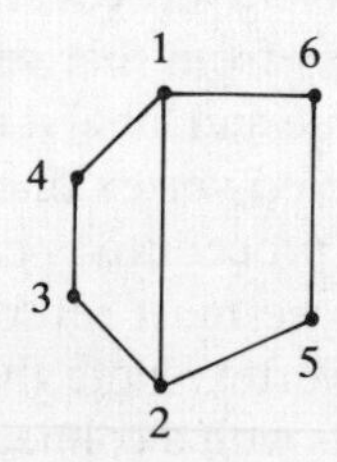

FIG. 2.4. *A labeling of vertices using the depth first search labeling procedure.*

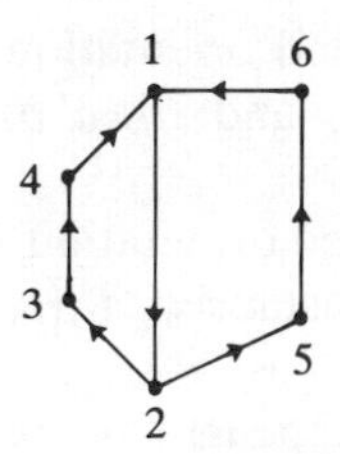

FIG. 2.5. *A strongly connected orientation from the depth first search labeling.*

$|E(G)|$ steps. By way of comparison, the first algorithm requires us to determine at the very first step whether or not u reaches v in D', a digraph with $|V(G)|$ vertices. This requires on the order of $|V(G)|^3$ computations (see Reingold, Nievergelt, and Deo (1977, p. 341)). Since $|E(G)|$ is on the order of $|V(G)|^2$, the first algorithm certainly is likely to be slower.

2.4. Efficiency. One of the problems with our discussion about moving traffic efficiently is that we have only been concerned with finding orientations which make it possible to get from one place to another, but have not been concerned with the possibility that it might become necessary to take long detours to do so. To make this discussion precise, let us define the *distance* $d_G(x, y)$ between two vertices x and y of a connected graph G to be the length of the shortest chain between them, and the distance $d_D(x, y)$ from vertex x to vertex y in a strongly connected digraph D to be the length of the shortest path from x to y. Note that $d_D(x, y)$ may not be the same as $d_D(y, x)$. Figure 2.6 shows two strongly connected orientations D and D' of a graph G. Notice that $d_D(a, b) = 11$, while $d_{D'}(a, b) = 3$. From the point of view of a person trying to get from a to b, D' is a much more efficient orientation. In general, one would like to obtain a strongly connected orientation of a graph G in which "on the whole," distances traveled are not too great. There are several ways to formulate this problem. Here are four formulations:

1) Find that strongly connected orientation D of G in which the average distance $d_D(a, b)$ over all a, b is as small as possible.

2) Find that strongly connected orientation D of G in which the maximum distance $d_D(a, b)$ over all a, b is minimized.

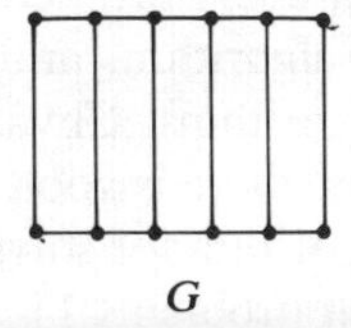
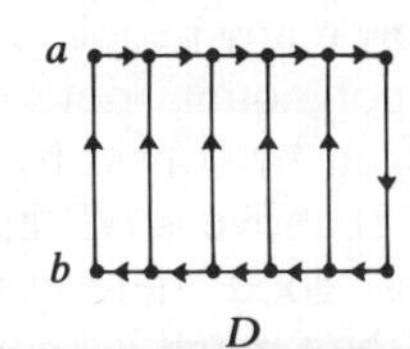

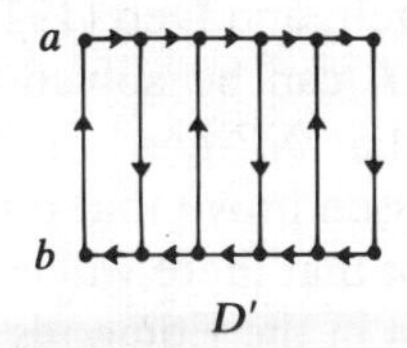

G D D'

FIG. 2.6. *Two different strongly connected orientations of G in which the distance from a to b differs.*

3) Find that strongly connected orientation D of G in which the difference between the distances $d_G(a, b)$ and $d_D(a, b)$ is on the average as small as possible.

4) Find that strongly connected orientation D of G in which the maximum of the differences between the distances $d_G(a, b)$ and $d_D(a, b)$ is as small as possible.

Chvátal and Thomassen (to appear) have recently obtained some very interesting (and somewhat discouraging) results about these problems. We define the *diameter* of a connected graph (strongly connected digraph) as the maximum distance between any two vertices. In particular, Chvátal and Thomassen prove that every connected, bridgeless graph of diameter d has a strongly connected orientation of diameter at most $2d^2+2d$. However, they show that the problem of finding that strongly connected orientation which minimizes diameter (problem 2)) is probably very difficult.

Let us be precise about that. Every finite problem has an algorithm for its solution: simply try all cases. However, beginning with the work of Edmonds (1965), algorists have searched for "good" procedures, i.e., procedures which will terminate in no more than $p(n)$ steps, where n is the size of the "input" and $p(n)$ is a polynomial. In studying such procedures, we shall distinguish between deterministic algorithms and nondeterministic algorithms. We shall make this distinction very informally. An algorithm can be thought of as passing from state to state. A *deterministic* algorithm may move to only one new state at a time, while a *nondeterministic* one may move to several new states at once. That is, a nondeterministic algorithm can explore several possibilities simultaneously. The class of problems for which there is a deterministic algorithm which terminates in polynomial time is called *P*. The class of problems for which there is a nondeterministic algorithm which terminates in polynomial time is called *NP*. An example of a problem we shall encounter which is in the class *NP* is the problem of determining whether a graph is colorable using a given number of colors. It is not known whether any problems in *NP* are in *P*, i.e., can be solved by a deterministic polynomial algorithm. We shall say a problem L is *NP-hard* if L has the following property: if L can be solved by a deterministic polynomial algorithm, then so can every problem in *NP*. A problem is *NP-complete* if it is *NP*-hard and it is in the class *NP*. Cook (1971) proved that there were *NP*-hard and *NP*-complete problems, and in particular his work implies that one problem we shall encounter below, that of finding the largest clique in a graph, is *NP*-hard. Karp (1972) showed that there were a great many *NP*-complete problems. For a good recent discussion of these notions, see Reingold, Nievergelt, and Deo (1977). Now many people doubt that every problem in the class *NP* can be solved by a polynomial deterministic algorithm, and hence doubt that *NP*-hard problems can be solved by such algorithms. Chvátal and Thomassen prove that problem 2) above is *NP*-hard. Thus, there is good reason to doubt that there will ever be a "good" deterministic algorithm for solving this problem in the Edmonds sense. Not much is known about problems 1), 3), and 4).

2.5. Inefficiency. Sometimes it is desirable to find a strongly connected orientation of a graph, but not to find one which is efficient in any of the distance-minimizing senses of the previous section. To give an example, the National Park Service has begun to take measures to discourage people from driving in the more heavily traveled sections of the park system. One approach has been to make (many) roads one way, but to do it inefficiently, i.e., so that long distances must be traveled to get from place to place. The reasoning is that if it is hard to get from place to place by car, people will consider other modes of transportation, such as bicycling, walking, or taking a bus. Roberts (1976a) discusses the one-way street network for the Yosemite Valley section of Yosemite National Park. Here, the diameter is much larger than the undirected diameter. In general, the problems of finding efficient orientations of a graph formulated in § 2.4 all have analogues for inefficient orientations. Good algorithms for finding such orientations are not known.

CHAPTER 3

Intersection Graphs

3.1. Transitive orientations. In what follows, it will be convenient to study types of orientations of a graph other than strongly connected ones. In particular, we shall be interested in orientations which lead to a transitive digraph and we shall ask what graphs have transitive orientations.

We say that a digraph D without loops is *transitive* if whenever there is an arc from u to v and an arc from v to w, and $u \neq w$, then there is an arc from u to w. In Fig. 3.1, digraphs D_1 and D_2 are transitive, while digraphs D_3 and D_4 are not, since the arc (u, w) is missing in the latter two cases. Digraph D_1 is a transitive

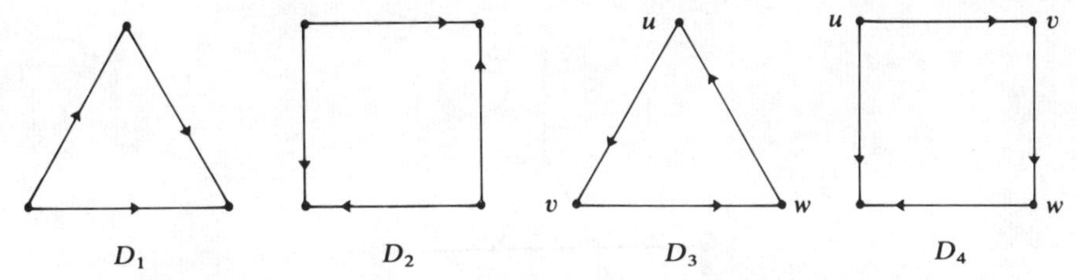

FIG. 3.1. *Digraphs D_1 and D_2 are transitive while D_3 and D_4 are not.*

orientation for the circuit of length 3, and digraph D_2 is a transitive orientation for the circuit of length 4. In general, let us ask whether the graph Z_n, the circuit of length n, has a transitive orientation. It is easy enough to see that Z_5 does not have such an orientation. For, if there were such an orientation, then by symmetry we could in Fig. 3.2 orient from a to b. Since there could be no arc from a to c after the orientation, edge $\{b, c\}$ would have to be oriented from c to b.

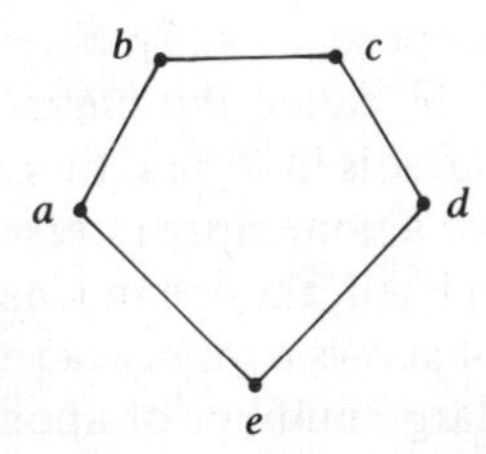

FIG. 3.2. *The graph Z_5.*

Similarly, edge $\{c, d\}$ would have to be oriented from c to d, edge $\{d, e\}$ from e to d, and edge $\{a, e\}$ from e to a. But now there would be arcs (e, a) and (a, b), and transitivity would imply the existence of an arc (e, b). This line of reasoning shows that Z_5 could have no transitive orientation. A similar argument shows that Z_n, for n odd and greater than 3, could have no such orientation. However, Z_n, for n even, does, in analogy to Z_4.

Another graph without a transitive orientation is the graph of Fig. 3.3. If there is such an orientation, then by symmetry we may assume that it goes from a to b. Transitivity now forces the following conclusions in the following order:

edge $\{b, e\}$ is oriented from e to b (or else there would have to be an arc from a to e),
edge $\{b, c\}$ is oriented from c to b,
edge $\{c, f\}$ is oriented from c to f,
edge $\{a, c\}$ is oriented from c to a.

Now, neither orientation on edge $\{a, d\}$ will suffice. For if the orientation goes from a to d, then we have arcs (c, a) and (a, d), but there could be no arc (c, d). If the orientation goes from d to a, then we have arcs (d, a) and (a, b), but there could be no arc (d, b).

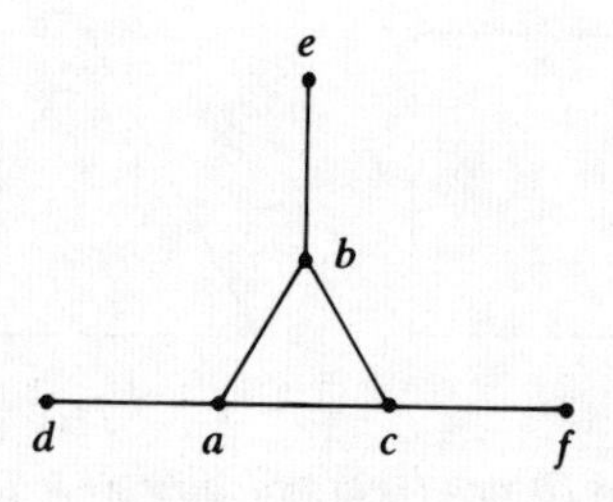

FIG. 3.3. *A graph which is not transitively orientable and not an interval graph but is a rigid circuit graph.*

Transitively orientable graphs were characterized by Ghouila-Houri (1962) and Gilmore and Hoffman (1964). Since we shall not need the characterization, we refer the reader to the literature for a discussion. We discuss the uniqueness of a transitive orientation in § 4.4.

3.2. Intersection graphs. Suppose $\mathcal{F} = \{S_1, S_2, \cdots, S_p\}$ is a family of sets.[2] We can associate a graph with $\mathcal{F}$, called the *intersection graph of* $\mathcal{F}$, as follows. The vertices of this graph are the sets in $\mathcal{F}$, and there is an edge between two sets S_i and S_j if and only if they have a nonempty intersection. (We shall as a general rule omit the loop from a set to itself, though in Chapter 4 it will be useful for us to include that loop.) Figure 3.4 shows a family of sets and its intersection graph. Intersection graphs arise in a large number of applications. A recent one is the following. Consider a collection of large corporations, for example the "Fortune 800." For the ith corporation, let S_i be the set consisting of members of the Board of Directors of this corporation. Levine (1976) and others study the

[2] We shall not find it necessary to require that the sets S_i be distinct.

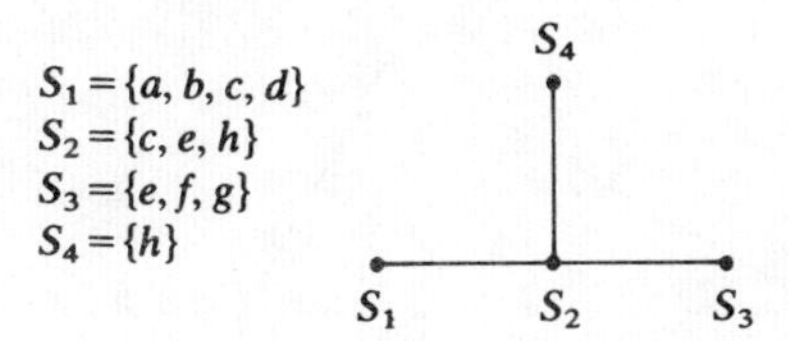

FIG. 3.4. *A family of sets and its intersection graph.*

intersection graph of the family of S_i in order to understand the network of corporate interlocks. (Incidentally, this graph has almost all its vertices (sets) in one connected component, and the median distance between two vertices in this component is 3.)

THEOREM 3.1 (Marczewski (1945)). *Every graph is (isomorphic to)*[3] *the intersection graph of some family of sets.*

Proof. Given G, let

$$S(u)=\{\{u, v\}: \{u, v\}\in E(G)\}\cup\{u\}.$$

It is easy to see that for all $u\neq v$ in $V(G)$,

$$\{u, v\}\in E(G)\Leftrightarrow S(u)\cap S(v)\neq\emptyset.$$

Thus, G is (isomorphic to) the intersection graph of the sets $S(u)$. Q.E.D.

Perhaps the most interesting applications of intersection graphs have arisen from taking special classes of sets. In the next section we consider such a class, the intervals on the real line.

3.3. Interval graphs and their applications. The intersection graph of a family of intervals on the real line[4] is called an *interval graph*. Given a graph G, we shall ask whether it is (isomorphic to) an interval graph. This is equivalent to the question: when is there an assignment of an interval $J(x)$ to each x in $V(G)$ so that for all $u\neq v$ in $V(G)$,

$$\{u, v\}\in E(G)\Leftrightarrow J(u)\cap J(v)\neq\emptyset. \tag{3.1}$$

(The intervals $J(x)$ do not have to be distinct, though it is easy to see that we may take them to be distinct without loss of generality.) Figure 3.5 shows an example of an interval graph. An *interval assignment* J satisfying (3.1) is obtained by taking $J(a)=(1, 5)$, $J(b)=(4, 8)$, $J(c)=(1, 14)$, $J(d)=(7, 11)$, and $J(e)=(10, 14)$.

Interval graphs arose from purely mathematical considerations (Hajos (1957)) and, independently, from a problem of genetics (Benzer (1959), (1962)). *Benzer's Problem* was the following. On the basis of mutation data, one can tell if

[3] Two graphs $G=(V, E)$ and $H=(W, F)$ are *isomorphic* if there is a one-to-one onto function $f: V\to W$ so that

$$\{u, v\}\in E(G)\Leftrightarrow\{f(u), f(v)\}\in E(H).$$

[4] The intervals may be open, closed, or half-open. However, without loss of generality, they may all be taken as open or all taken as closed.

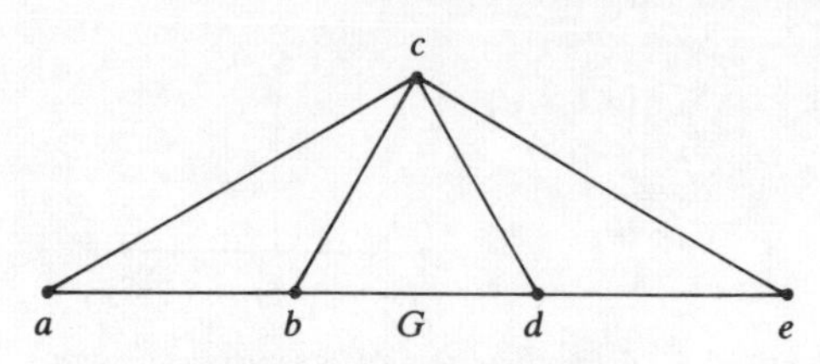

FIG. 3.5. *An interval graph.*

two subsets of the fine structure inside the gene overlap. Is this overlap information consistent with the hypothesis that the fine structure inside the gene is linear? It is if the graph defined by the overlap information is an interval graph. (For a more detailed discussion, see Roberts (1976a).)

Interval graphs also arise in the measurement of preference and indifference. Suppose we have a range of possible values of an item. If we are trying to decide between two different items, we might reasonably be expected to prefer item u to item v if and only if the interval of values $J(u)$ of item u is strictly to the right of the corresponding interval $J(v)$, i.e., if and only if every element of $J(u)$ is larger than every element of $J(v)$. Similarly, we might reasonably be expected to be indifferent between u and v if and only if the intervals $J(u)$ and $J(v)$ overlap. If this is the case, indifference should define an interval graph. We shall return to the study of indifference in § 4.1.

In seriation in the social sciences, we try to put a collection of items or objects in some serial order or sequence. For example, in archaeology we are interested in sequence dating a collection of artifacts. In psychology, we want to put some traits in a developmental order, or order individuals according to their opinions. In political science, we sometimes wish to order political candidates from liberal to conservative. One approach to seriation is to start with overlap information. For example, we ask in archaeology whether or not the time intervals during which two artifacts existed overlapped. We obtain this information from observation of graves. We seek an assignment of time intervals so that artifact a and artifact b were found in common in some grave if and only if the time interval associated with a overlaps with the time interval associated with b. Such an assignment can be obtained if and only if the "found in common in some grave" graph is an interval graph. The intervals are a *possible chronological order.* They can, unfortunately, differ significantly from the "real chronological order." We shall discuss how in some detail in §4.4. This approach to seriation in archaeology is carried out in Kendall (1963), (1969a,b). A similar approach in developmental psychology is carried out by Coombs and Smith (1973). We shall return to a discussion of seriation in §4.2.

We shall also discuss applications of interval graphs to the phasing of traffic lights, to the assignment of mobile radio telephone frequencies, and to the study of ecological phase space. For now, we turn to the problem of characterizing interval graphs.

3.4. Characterization of interval graphs. Let us begin by observing that Z_3, the circuit of length 3, is an interval graph, but that Z_n for $n \geqq 4$ is not. The former we leave to the reader. To see the latter, let us consider the case of Z_4. Let the vertices of Z_4 be labeled as in Fig. 3.6. If there were an interval assignment J satisfying (3.1), then $J(a)$ and $J(b)$ would have to overlap, since there is an edge between a and b. Without loss of generality, $J(b)$ is a bit to the right of $J(a)$. ($J(b)$ cannot completely lie inside $J(a)$ for otherwise we could not have $J(c)$ overlapping $J(b)$ without overlapping $J(a)$, as is required. Similarly, $J(a)$ cannot lie inside $J(b)$.) Now $J(c)$ overlaps $J(b)$ but not $J(a)$, so $J(c)$ must be as pictured in Fig. 3.6. Finally, $J(d)$ must overlap both $J(a)$ and $J(c)$, but not $J(b)$. Where could $J(d)$ be? A similar argument holds for any $n \geqq 4$.

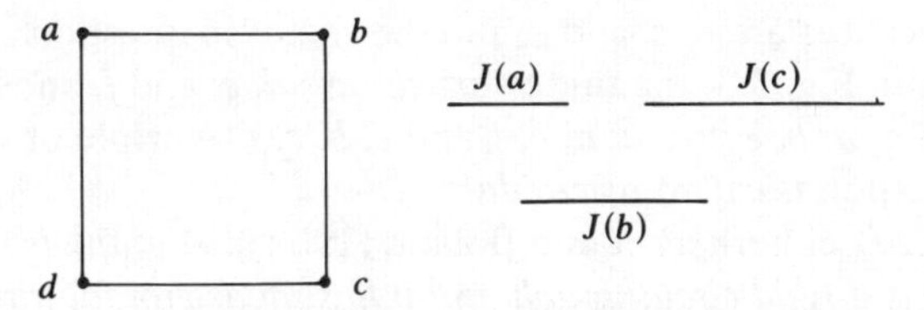

FIG. 3.6. *The argument that Z_4 is not an interval graph.*

A graph $H = (W, F)$ is a *subgraph* of a graph $G = (V, E)$ if $W \subseteq V$ and $F \subseteq E$. (Notice that by saying H is a graph, we imply that F is a set of pairs from W.) H is a *generated subgraph* if F consists of all edges from E joining vertices in W. It is easy to see that if G is an interval graph, then every generated subgraph must also be an interval graph. However, this is not the case for every subgraph. Thus, if G is an interval graph, it has the property that no graph Z_n, $n \geqq 4$, is a generated subgraph. A graph G with this property is called a *rigid circuit graph*, or a *triangulated graph.* Rigid circuit graphs have the property that whenever $x_1, x_2, \cdots, x_t, x_1$ is a circuit of length $t \geqq 4$, then there is in the graph a *chord*, an edge of the form $\{x_i, x_j\}$, where $j \neq i \pm 1$ and addition is considered modulo t. The graph of Fig. 3.7 is an example of a rigid circuit graph. Although there are circuits of length 4 or greater, every such circuit has a chord. This graph is also an example of a rigid circuit graph which is not an interval graph. To see that, note that if there were an interval assignment J, then $J(a)$, $J(c)$, and $J(e)$ would have

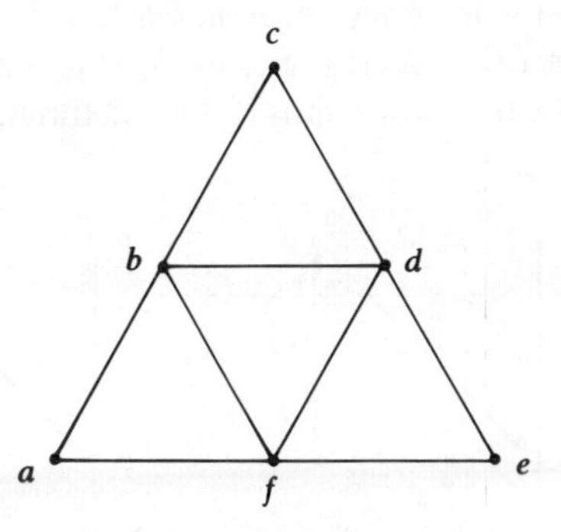

FIG. 3.7. *A rigid circuit graph which is not an interval graph.*

to be pairwise disjoint intervals. One of them, say $J(c)$ without loss of generality, would have to be in between the other two. Then $J(f)$ would have to overlap $J(a)$ and $J(e)$ but not $J(c)$, which is impossible. The graph of Fig. 3.3 is another example of a graph which is a rigid circuit graph but not an interval graph. Verification of this fact is left to the reader.

What these two examples of noninterval graph rigid circuit graphs have in common is a triple of vertices x, y, and z with the property that there are chains C_{xy} between x and y, C_{xz} between x and z, and C_{yz} between y and z with x not *adjacent* (joined by an edge) to any vertex of C_{yz}, y not adjacent to any vertex of C_{xz}, and z not adjacent to any vertex of C_{xy}. For example, in the graph of Fig. 3.7, the three vertices are a, c, and e, and the chains in question are a, b, c, and a, f, e, and c, d, e. In Fig. 3.3, the three vertices are d, e, and f, and the three chains in question are d, a, b, e and d, a, c, f and e, b, c, f. A triple of vertices with the property in question is called *asteroidal.*

THEOREM 3.2 (Lekkerkerker and Boland (1962)). *A graph is an interval graph if and only if it is a rigid circuit graph and it has no asteroidal triple.*

To give yet a second characterization of interval graphs, let us define the *complement* G^c of the graph G to be a graph with the same vertex set as G, but with an edge between $x \neq y$ if and only if there is no edge between x and y in G. Now suppose G is an interval graph and J is an interval assignment for G. Then we can define an orientation on G^c as follows: orient the edge $\{x, y\}$ of G^c from x to y if and only if $J(x)$ is strictly to the right of $J(y)$. This orientation is well defined because the intervals $J(x)$ and $J(y)$ corresponding to the vertices x and y do not overlap. It is clear that the orientation so defined must be transitive, for the relation "strictly to the right of" on a set of intervals is transitive. Hence, we have shown that if G is an interval graph, then G^c is transitively orientable.

THEOREM 3.3 (Gilmore and Hoffman (1964)). *A graph G is an interval graph if and only if Z_4 is not a generated subgraph and G^c is transitively orientable.*

To illustrate this theorem, we observe that the graph of Fig. 3.5 does not have Z_4 as a generated subgraph. Moreover, the complementary graph, which is shown in Fig. 3.8, is transitively orientable. A transitive orientation is also shown in that figure.

We have already seen the necessity of the conditions in Theorem 3.3. We briefly sketch a proof of sufficiency. A *complete graph* is a graph in which every vertex is joined to every other vertex, and a *clique* in a graph is a subgraph which is complete. A clique is called *dominant* if it is maximal, i.e., if it is not contained

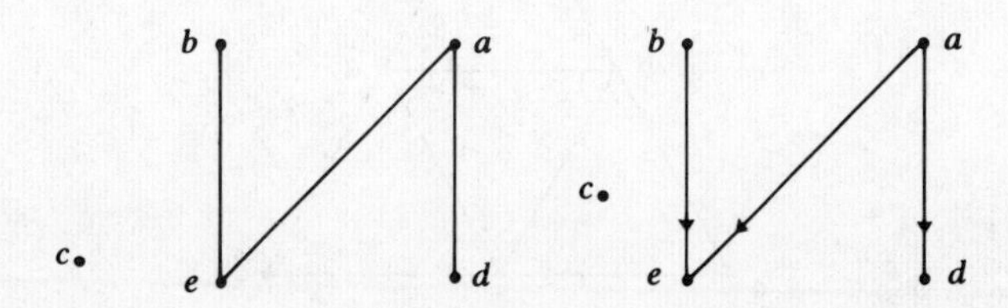

FIG. 3.8. *The complement of the graph of Figure 3.5 and a transitive orientation for the complement.*

(when considered as a set of vertices) in any larger clique. Suppose we are given a graph G which does not have Z_4 as a generated subgraph and suppose we are given a transitive orientation for G^c. Let $\mathscr{C}$ be the collection of dominant cliques of G.[5] We can use the orientation of G^c to order the cliques in $\mathscr{C}$. In particular, we take a clique K before a clique L if whenever u is in K and v is in L and $\{u, v\} \notin E(G)$, then the edge $\{u, v\}$ of G^c is oriented from u to v. One needs to prove that this ordering of dominant cliques is well defined, by first showing that there always are such u and v for every $K \neq L$ and then showing that a different u' from K and v' from L cannot lead to the opposite ordering. The details are left to the reader. The proof uses the hypothesis that G does not have Z_4 as a generated subgraph. To illustrate the procedure, consider the graph G of Fig. 3.5 and the transitive orientation of the complement which is shown in Fig. 3.8. The dominant cliques of G are $K = \{a, b, c\}$, $L = \{b, c, d\}$ and $M = \{c, d, e\}$. Since $a \in K$ and $d \in L$ and there is no edge in G between a and d, the orientation a to d in G^c tells us to take K before L. Similarly using vertices b and e, we see that L comes before M, and using vertices a and e, we see that K comes before M, so the ordering of dominant cliques is K, L, M. It is not hard to show, using the transitivity of the orientation of G^c, that one always gets a linear ordering $K_1, K_2, \cdots, K_p$ of dominant cliques. Moreover, one sees that this ordering of the dominant cliques has the following property P: if $\alpha < \beta < \gamma$ and a vertex a belongs to K_α and K_γ, then it also must belong to K_β. Hence, we may take $J(a)$ to be the interval $[\alpha, \gamma]$, where K_α is the first dominant clique in the ordering to which a belongs, and K_γ is the last dominant clique in the ordering to which a belongs. In our example, $K = K_1$, $L = K_2$, and $M = K_3$. Since a belongs only to K_1, we obtain $J(a) = [1, 1]$, an interval consisting of a single point. Similarly, $J(b) = [1, 2]$, $J(c) = [1, 3]$, $J(d) = [2, 3]$ and $J(e) = [3, 3]$. It is easy to see that this assignment of intervals satisfies (3.1). For a more detailed discussion of why this procedure works in general, see Roberts (1976a).

An ordering of dominant cliques which satisfies property P above is called *consecutive*. G is an interval graph if and only if there is an ordering of the dominant cliques of G which is consecutive. For if there is such an ordering, then the above assignment J shows that G is an interval graph. Conversely, if G is an interval graph, then G does not have Z_4 as a generated subgraph and G^c has a transitive orientation, so the above construction using $\mathscr{C}$ shows that there is an ordering of the dominant cliques of G which is consecutive. It is convenient to restate this result in terms of matrices. If G is any graph, its *dominant clique-vertex incidence matrix* M is defined as follows. The rows of M correspond to the dominant cliques, and the columns to the vertices. The entry m_{ij} is 1 if the jth vertex belongs to the ith dominant clique, and it is 0 otherwise. If there is an ordering of dominant cliques which is consecutive, the corresponding ordering of rows of M gives rise to a matrix with the 1's in each column appearing consecutively. We say that a matrix A of 0's and 1's has the *consecutive* 1'*s property* (*for*

[5] Of course, as we have observed, even identification of the largest clique of a graph is an *NP*-hard problem. Hence, it is not in general easy to identify the collection $\mathscr{C}$.

columns) if it is possible to permute the rows so that the 1's in each column appear consecutively.

THEOREM 3.4 (Fulkerson and Gross (1965)). *A graph G is an interval graph if and only if its dominant clique-vertex incidence matrix M has the consecutive* 1's *property*.

To illustrate this theorem, let us note that the following matrix is the dominant clique-vertex incidence matrix for the graph Z_4 with the vertex labeling of Fig. 3.6. There is no permutation of the rows so that the 1's in each column appear consecutively.

$$M = \begin{array}{c} \\ ab \\ bc \\ cd \\ ad \end{array} \begin{array}{c} \begin{array}{cccc} a & b & c & d \end{array} \\ \begin{pmatrix} 1 & 1 & 0 & 0 \\ 0 & 1 & 1 & 0 \\ 0 & 0 & 1 & 1 \\ 1 & 0 & 0 & 1 \end{pmatrix} \end{array}.$$

3.5. Circular arc graphs. Various families of geometric interest give rise to useful classes of intersection graphs. In § 5.1, we shall consider rectangles in the plane, and more generally "boxes" in n-space. In this section, let us briefly consider the graphs which are intersection graphs of arcs on a given circle, the so-called *circular arc graphs*. It is easy to see that every interval graph is a circular arc graph, but that the converse is false: Z_4 is a counterexample. However, Z_4 together with an isolated additional vertex is not a circular arc graph as the reader can readily verify. It is easy to see that a graph G is a circular arc graph if and only if its dominant clique-vertex incidence matrix M has the *circular* 1'*s property*, i.e., there is a permutation of the rows so that the 1's in each column appear consecutively if 1's are allowed to continue from bottom to top. Circular arc graphs have been studied in Tucker (1970), (1971).

3.6. Phasing traffic lights. In this section we shall discuss the application of interval graphs and circular arc graphs to the phasing of traffic lights. The goals of traffic light phasing are to have traffic move safely and efficiently. With increasing concern about energy use, the latter goal is becoming of increased importance. Consider a traffic intersection at which we wish to install a new traffic light. Approaching the traffic intersection are various *traffic streams*, patterns or routes through the intersection which traffic takes. Figure 3.9 shows a traffic intersection with several traffic streams labeled with the letters a through f. The intersection has a two-way street meeting a one-way street. Certain traffic streams are judged to be compatible with each other, in the sense that they can be moving at the same time without dangerous consequences. The decision about compatibility is made ahead of time, by a traffic engineer, and may be based on estimated volume of traffic in a stream as well as the traffic pattern. The compatibility information can be summarized in a graph G, the *compatibility graph*. The vertices of G are the traffic streams, and two streams are joined by an edge if and only if they are judged compatible. Figure 3.9 shows such a compatibility graph for the traffic intersection in question. Notice that, for example, stream c, the left-turning traffic, is judged compatible with stream f, the straight and right-turning traffic in the other street, but not with stream e, the left-turning traffic in the other street. In traffic light phasing, we wish to assign a period of time to each stream during which it receives a green light, and to do it in such a way that only compatible traffic streams can get green lights at the same time. There is a cycle of green and red lights, and then after the cycle is finished, it begins again, over and over.

We may think of the time during the cycle as being kept on a large clock, and the time during which a given traffic stream gets a green light corresponds to an arc on the circumference of the clock circle. (We assume that a given stream receives only one continuous green light during each cycle.) Then a *feasible green light assignment* consists of an assignment of an arc of the circle to each traffic stream so that only compatible traffic streams are allowed to receive overlapping arcs. In terms of the compatibility graph, only vertices joined by an edge are allowed to receive overlapping arcs. This is not the same as the intersection graph idea, as we are not forcing compatible vertices to get overlapping arcs. However, if we look at the intersection graph corresponding to any feasible green light assignment, this will be a subgraph of the compatibility graph. It is not necessarily a generated subgraph. Indeed, it corresponds to a subgraph with the same vertex set as the compatibility graph, but one with perhaps some of the edges deleted, a so-called *spanning subgraph.* Figure 3.9 shows such a feasible green light assignment and its corresponding intersection graph H. Note that H must be a circular arc graph. Thus, feasible green light assignments correspond to spanning subgraphs of G which are circular arc graphs. It is not necessarily the case that G itself is a circular arc graph, although in this case it is. If we require that no green light time period overlap a starting time, i.e., that a cycle begins with all red lights, then the intersection graph corresponding to a feasible green light assignment is a spanning subgraph of G which is an interval graph. Notice that the compatibility graph G of Fig. 3.9 is not an interval graph, since the subgraph generated by vertices b, c, e, and f is Z_4. However, the spanning subgraph H is an interval graph.

Some feasible green light assignments are very uninteresting. For example, we can assign to each traffic stream an empty green arc and obtain a feasible assignment. What makes one feasible assignment better than another? We usually have in mind some criterion. For example, we might wish to minimize the total amount of waiting time, i.e., the total amount of red light time in a cycle. Or we might wish to minimize a weighted sum of red light times by weighting more heavily the red light time for heavily traveled traffic streams. Or, as Stoffers (1968) points out, we might have some information about expected arrival times of different traffic streams, and we might wish to penalize starting times for being far from the traffic stream's expected arrival time and minimize the penalties.

Let us illustrate the procedure for finding an optimal green light assignment if the criterion is to minimize the total red light times. We shall also make the

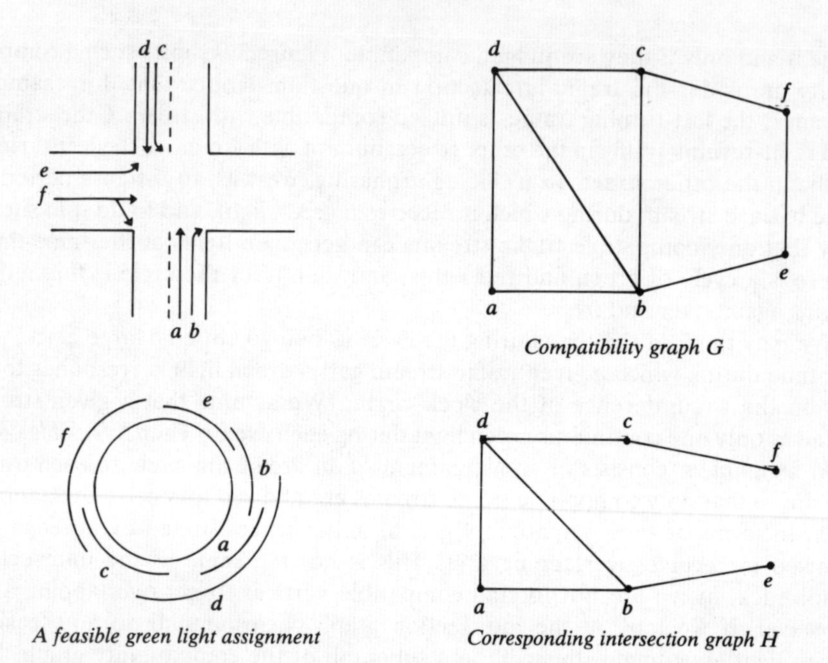

A feasible green light assignment

Corresponding intersection graph H

FIG. 3.9. *A traffic intersection, its compatibility graph, a feasible green light assignment, and the corresponding intersection graph.*

assumption that each green light arc must be a certain minimal length. We then follow Stoffers' (1968) procedure. For each circular arc graph spanning subgraph of G, or for each interval graph spanning subgraph if we are willing to make the special assumption about all red lights at the starting time, let us generate corresponding feasible green light assignments by considering orderings of dominant cliques which are consecutive. For concreteness, we handle the interval graph case. Let $K_1, K_2, \cdots, K_p$ be a consecutive ordering of dominant cliques of interval graph spanning subgraph H. Each clique K_i corresponds to a *phase* during which all streams in that clique get green lights. We then start a given traffic stream off with green during the first phase in which it appears, and keep it green until the last phase in which it appears. By consecutiveness of the ordering, this leads to an arc of the clock circle. In our example, the graph H has as one consecutive ordering of dominant cliques the ordering $K_1=\{e, b\}$, $K_2=\{b, a, d\}$, $K_3=\{d, c\}$, $K_4=\{c, f\}$. Thus, there are four phases. In phase 1, traffic streams e and b get green lights (the one-way street turns left and one right-turn light is on), then in phase 2 streams b, a, d get green lights (the left-turn light is turned off, and the north-south traffic starts up), and so on. Suppose we assign to each clique K_i a duration d_i. What should the durations d_i be so that the sum of the red light times is as small as possible? The answer is obtained by observing the following: a gets a red light during phases K_1, K_3, and K_4, so a's total red light time is $d_1+d_3+d_4$. Similarly, b's red light time is d_3+d_4. The total red light time for all traffic streams is given by

$$(d_1+d_3+d_4)+(d_3+d_4)+(d_1+d_2)+(d_1+d_4)+(d_2+d_3+d_4)+(d_1+d_2+d_3),$$

which equals

$$4d_1+3d_2+4d_3+4d_4. \tag{3.2}$$

If the minimum green light time for a stream is 20 seconds, and the total cycle is 120 seconds, we wish to minimize (3.2) subject to the constraints that each $d_i \geqq 0$ and

$$d_2 \geqq 20, \tag{3.3}$$

$$d_1+d_2 \geqq 20, \tag{3.4}$$

$$d_3+d_4 \geqq 20, \tag{3.5}$$

$$d_2+d_3 \geqq 20, \tag{3.6}$$

$$d_1 \geqq 20, \tag{3.7}$$

$$d_4 \geqq 20, \tag{3.8}$$

$$d_1+d_2+d_3+d_4=120. \tag{3.9}$$

Constraints (3.3) through (3.8) correspond to the statements that *a* through *f*, respectively, receive green light times of at least 20 seconds. The solution to our linear programming problem is easy in our case. By (3.9), minimizing (3.2) is equivalent to minimizing $d_1+d_3+d_4$. Since d_1 and d_4 must be at least 20 and d_3 at least 0, it is clear that (3.2) is minimized by taking $d_1=d_4=20$, $d_3=0$, and $d_2=80$.

To find an optimal feasible green light assignment, we must identify each interval graph (or circular arc graph) spanning subgraph *H* of *G*. For each, we must find all the different consecutive (circular) orderings of dominant cliques and for each such ordering, we must find an optimal solution of phase durations.[6] Then we can put all this together to find an optimal solution for the entire graph. The reader might find it enlightening to consider in our example or examples of his own choosing alternative interval graph spanning subgraphs and alternative phasings which arise from different consecutive orderings of their dominant cliques.

3.7. The mobile radio frequency assignment problem. Mobile radio telephone systems, such as those assigned to police cruisers, operate in different zones. Each zone receives a band of frequencies which can be used within it. These bands are often intervals, though more generally they are unions of intervals. The mobile telephones in one zone can cause interference with those in another zone. In that case, their bands of frequencies should not overlap. In

[6] In general, as we have remarked, even identifying all the dominant cliques involves a lengthy computation for larger graphs. However, we are here dealing with relatively small graphs.

assigning bands to zones, one wants to take these conflicts into account, and also meet certain requirements on minimal bandwidth for given zones.

Following Gilbert (1972), we can formulate this problem graph-theoretically. The vertices of a *conflict graph* are the zones, and two zones are adjacent if and only if they conflict. Then, we wish to assign to each zone i a band $B(i)$—let us say an interval—so that if there is an edge between zones i and j, then we have $B(i) \cap B(j) = \varnothing$. Moreover, we wish to do this so that each $B(j)$ has certain minimal length.

Looked at this way, we see that this problem is reducible to the traffic light phasing problem. We simply consider the feasible green light assignment problem on the complementary graph of the conflict graph. We shall have more to say about the mobile radio frequency assignment problem in § 6.5.

Notes added in press. Both the radio frequency assignment problem and the traffic light phasing problem are of interest for unions of intervals, not just for intervals. Recent work on intersection graphs of unions of intervals can be found in Griggs and West (1977) and Trotter and Harary (1977). Recent results on the radio frequency assignment and traffic light phasing problems for sets other than intervals and unions of intervals and on the duality between these two problems can be found in Roberts (to appear).

CHAPTER 4

Indifference, Measurement, and Seriation

4.1. Indifference graphs. Many decisions to be made by individuals, groups, or by society as a whole require the ability to measure variables which are not as easy to measure as physical variables like temperature or mass. The need to measure such things as preference, aesthetic appeal, agreement, and so on has given rise to a theory of measurement which covers social scientific variables as well as physical variables. One of the goals of measurement is to organize data into some coherent structure, so that the underlying patterns can be pinpointed. Techniques of measurement and scaling are used often in this way in the social sciences, in particular in trying to understand opinions, viewpoints, and the like. The resulting scales have uses in a variety of decisions which need to be made by individuals, groups, or societies. In this section, we shall discuss the use of graph-theoretical tools in one measurement problem, the measurement of indifference. We then turn to a related problem, that of seriation of data, and show the connection with the results on indifference measurement.

Suppose an individual expresses preference among alternatives in a set V. We write $x\mathrm{P}y$ to mean that x is preferred to y. If an individual is not forced to choose between x and y, he is allowed to be indifferent. (He is indifferent between x and y if and only if he prefers neither.) We write $x\mathrm{I}y$ to mean that he is indifferent between x and y. Measurement of preference corresponds to the assignment of numbers which "preserve" the expressed preferences. In particular, the goal is to assign a real number $f(x)$ to each x in V so that for all x, y,

$$(4.1) \qquad x\mathrm{P}y \Leftrightarrow f(x) > f(y),$$

i.e., x is preferred to y if and only if x receives a higher number than y. If it is possible to obtain such a function f, then indifference corresponds to equality:

$$(4.2) \qquad x\mathrm{I}y \Leftrightarrow f(x) = f(y).$$

Equation (4.2) implies that indifference is transitive. The economist Armstrong (1939), (1948), (1950), (1951) was among the first to argue that indifference is not transitive. (Menger (1951) claims that such arguments go back at least to Poincaré.) One argument against the transitivity of indifference is the following argument of Luce (1956). We are almost certainly not indifferent between a cup of coffee with no sugar and a cup with five spoons of sugar. However, if we add sugar to the first cup one grain at a time, we will almost certainly be indifferent between successive cups. Transitivity of indifference would imply that we are indifferent between the cup without sugar and the cup with five spoons of sugar.

Examples such as this suggest that indifference does not correspond to equality, but rather to closeness. Speaking precisely, suppose δ is a positive number measuring closeness. Then we might seek an assignment of numbers $f(x)$ to elements of V so that for all x, y in V,

$$xIy \Leftrightarrow |f(x)-f(y)|<\delta. \tag{4.3}$$

Under what circumstances does there exist such an assignment? The answer to this question is important in measurement, because it tells us when a particular way of measuring or scaling data (judgments of indifference) can be carried out.

To answer this question, we translate it into a graph-theoretic one. We let V be the set of vertices of a graph G, and we draw an edge between x and y if and only if xIy. Notice that the resulting graph has a loop at each vertex. In the rest of this chapter, it will be convenient to allow loops in our graphs. In this section, we will assume that every graph has a loop at each vertex. We would like to assign a number $f(x)$ to each vertex of the graph G so that vertices joined by an edge get numbers which are close (within δ) and vertices not joined by an edge get numbers which are not close (not within δ). If there is such an assignment of numbers, we say that G is an *indifference graph*. It is easy to give examples of indifference graphs. Consider for instance the graph of Fig. 4.1.[7] If $\delta=1$, an assignment of numbers satisfying (4.3) is given by $f(a)=0$, $f(b)=.3$, $f(c)=.7$, $f(d)=1.1$, and $f(e)=1.4$. (Incidentally, it is clear that there is an assignment f with $\delta=1$ if and only if there is an assignment f for any positive δ.) To give an example of a graph which is not an indifference graph, consider Z_4 with the labeling of vertices given in Fig. 4.2.[8] If there is a function f satisfying (4.3), then without loss of generality $f(b)\geqq f(a)$. Now since a and b are joined by an edge and similarly b and c, the numbers $f(a)$ and $f(b)$ must be within δ and so must the numbers $f(b)$ and $f(c)$. However, a and c are not joined by an edge, so the

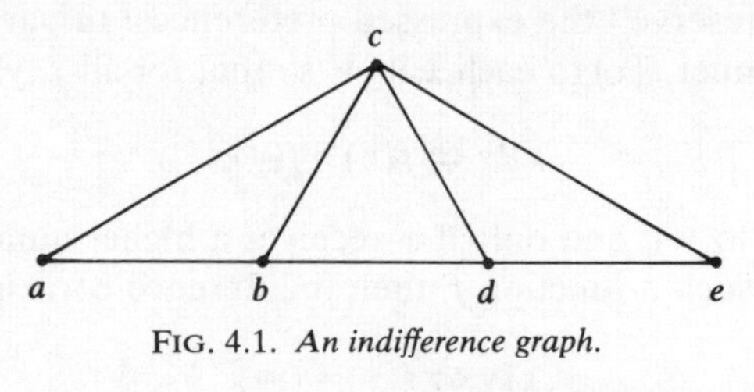

FIG. 4.1. *An indifference graph.*

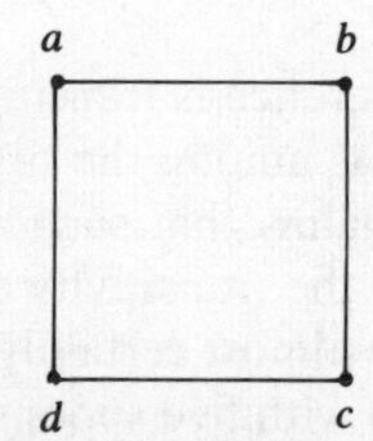

FIG. 4.2. *The graph Z_4.*

[7] We shall suppress the loop at each vertex in this and all graphs of this section.

[8] Again, there is a (suppressed) loop at each vertex.

numbers $f(a)$ and $f(c)$ must be at least δ apart. It follows that these numbers are

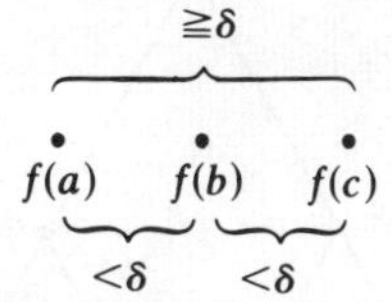

$\geq\delta$

$f(a)$ $f(b)$ $f(c)$

$<\delta$ $<\delta$

FIG. 4.3. *The argument that Z_4 is not an indifference graph.*

arranged as in Fig. 4.3. Now $f(d)$ must be within δ of $f(a)$ and $f(c)$, but not within δ of $f(b)$. This is impossible. Thus, Z_4 is not an indifference graph. Neither is Z_n, any $n \geqq 4$.

It is also easy to see that Z_n, $n \geqq 4$, is not an indifference graph by considering the intervals

(4.4) $$J(a) = (f(a) - \delta/2, f(a) + \delta/2).$$

Now, if f satisfies (4.3), then

$$xIy \Leftrightarrow J(x) \cap J(y) \neq \varnothing.$$

Hence, the graph of indifferences is an interval graph.[9] In particular, the graph of indifferences cannot be Z_n, $n \geqq 4$, or indeed have such a Z_n as a generated subgraph. The noninterval graphs of Figs. 3.3 and 3.7, which are repeated in Figs. 4.4 and 4.5, give other examples of nonindifference graphs. It is instructive for the reader to try to argue this directly. Not every interval graph is an indifference graph. Consider the graph of Fig. 4.6. Suppose there were a function f satisfying (4.3). Without loss of generality we have $f(a) < f(b) < f(c)$. Since there are no edges among a, b, and c, we have $f(a)$ and $f(c)$ at least 2δ apart. However, $f(d)$ must be within δ of both $f(a)$ and $f(c)$, which is impossible.[10]

It is easy to show that if G is an indifference graph, then so is every generated subgraph. Hence, we have shown one direction of the following theorem.

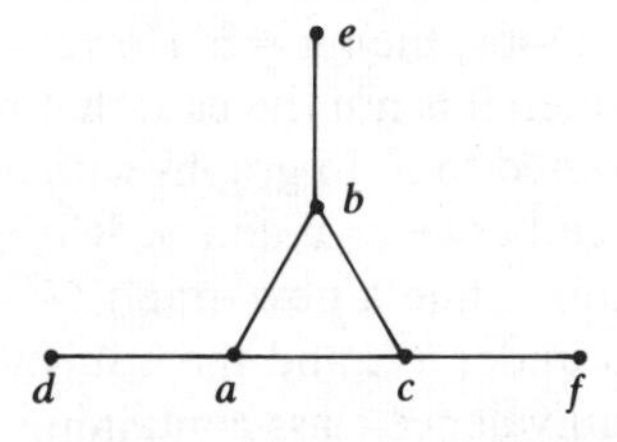

FIG. 4.4. *A graph which is not an indifference graph.*

[9] Every interval $J(a)$ in (4.4) has the same length. We say G is a *unit interval graph* if it is the intersection graph of a family of (open) intervals of the same length (without loss of generality unity). The indifference graphs are exactly the unit interval graphs.

[10] In a sense, the graph of Fig. 4.6 is the only interval graph which is not an indifference graph. In Roberts (1969a), it is shown that if G is an interval graph, then G is an indifference graph if and only if the graph of Fig. 4.6 is not a generated subgraph of G.

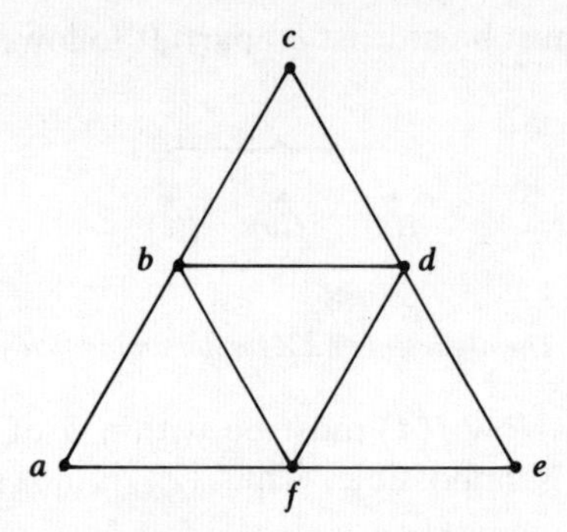

FIG. 4.5. *A graph which is not an indifference graph.*

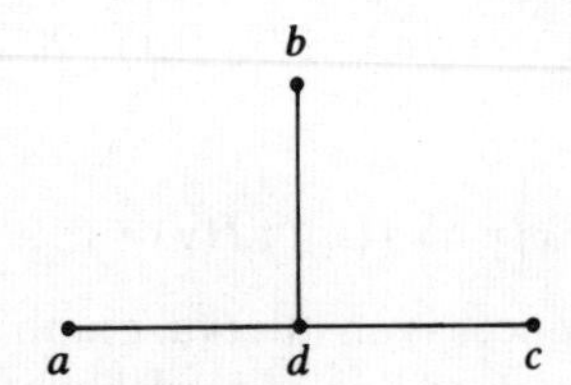

FIG. 4.6. *An interval graph which is not an indifference graph.*

THEOREM 4.1 (Roberts (1969a)). *A graph G (with a loop at each vertex) is an indifference graph if and only if it is a rigid circuit graph and it does not have any of the graphs of Figs.* 4.4, 4.5, *and* 4.6 *as a generated subgraph.*

This theorem is an example of a *forbidden subgraph characterization*, a theorem which characterizes a class of graphs by telling which configurations cannot appear as generated subgraphs.

To gain a little more insight into indifference graphs, let us define a relation $\approx$ on the set of vertices of a graph by taking $x \approx y$ if and only if x and y are joined by edges to exactly the same vertices. In particular, if the graph of Fig. 4.6 is considered a graph without loops, then $a \approx b$. However, if we consider it a graph with a loop at each vertex, then it is not the case that $a \approx b$, since a is joined to a by an edge while b is not joined to a. In graphs with a loop at each vertex, $x \approx y$ implies that there is an edge between x and y. It is easy to show that $\approx$ is an equivalence relation. We can define a new graph $G/\approx$ as follows. The vertices are the equivalence classes under $\approx$, and the equivalence class containing x is joined by an edge to the equivalence class containing y if and only if x is joined to y. This is clearly well defined. We say that G is *reduced* if G is isomorphic to $G/\approx$. We shall use this notion of equivalence and the idea of reduction to give another characterization of the indifference graphs.

Consider a finite set of points on the real line. There are two end points. We shall try to capture by the following definition what vertices of a graph can be mapped into end points under a function f satisfying (4.3). We say that vertex x of graph G is an *extreme vertex* if whenever there are edges from x to y and to z, then y and z are joined by an edge and, moreover, there is a vertex w so that y and z are joined to w but x is not joined to w. In the graph of Fig. 4.5, vertex a is an extreme vertex. For the only pair of vertices to which it is joined are b and f, b and f are joined to each other, and there is a vertex d joined to b and f but not to a. Similarly, c and e are extreme vertices here. In the graph of Fig. 4.4, vertices labeled d, e, and f are vacuously extreme vertices. In the graph of Fig. 4.6, vertices a, b, and c are extreme vertices. However, in the graph Z_n, $n \geqq 4$, there are no extreme vertices. For every vertex is joined to two others which are not joined.

THEOREM 4.2 (Roberts (1969a)). *A graph G (with a loop at each vertex) is an indifference graph if and only if for every connected generated subgraph H of G, $H/\approx$*[11] *is either a single vertex or has exactly two extreme vertices.*

This theorem explains why the graphs Z_n, $n \geqq 4$, and the graphs of Figs. 4.4, 4.5, and 4.6 are not indifference graphs. If each of these is considered a graph with loops at each vertex, then they are each connected, each is already reduced, and each has either too few or too many extreme vertices. It is necessary to deal with the reduction, because if there are equivalent vertices, then a connected indifference graph can have fewer than two extreme vertices if our present definition of extreme vertex is used—consider the complete graph on 3 vertices. Also, it is not sufficient to state Theorem 4.2 in terms of connected components H. The graph of Fig. 4.7 shows why. It is connected and reduced and has just two extreme vertices (a and e). However, the subgraph generated by vertices x, a, c, and e is connected and reduced and has three extreme vertices (a, c, and e)—this subgraph is isomorphic to the graph of Fig. 4.6.

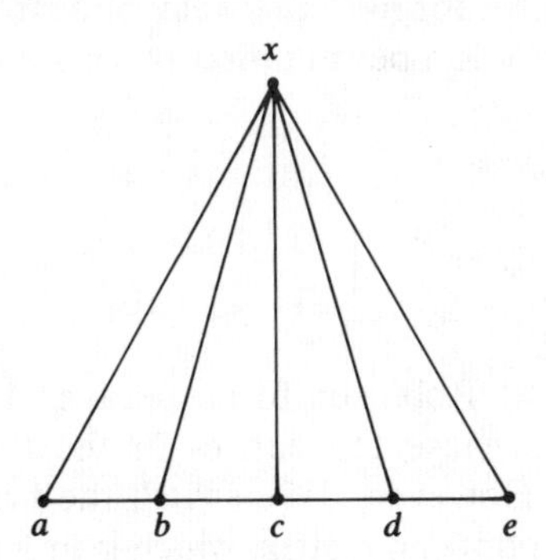

FIG. 4.7. *A connected, reduced graph which is not an indifference graph but has only two extreme vertices, a and e.*

4.2. Seriation. An important problem of data analysis in the social sciences involves the sequencing of objects in some order. As we have pointed out earlier, archaeologists try to order artifacts by age, psychologists try to order subjects along steps of development or shades of opinion, and political scientists try to order politicians from liberal to conservative. In making decisions, we

[11] The relation $\approx$ is the one defined from H.

sometimes like to order alternatives from least risky to most risky, from least conservative to most conservative, etc. One approach to seriation begins with a measure r_{ij} of the similarity of alternatives i and j, with r_{ij} higher than r_{kl} if i and j are more similar than k and l. We shall assume that $R=(r_{ij})$ is a symmetric matrix and discuss how to obtain sequences of objects from R.

We say that matrix R is in *strong Robinson form* if whenever $i \leqq j \leqq k \leqq l$, then $r_{jk} \geqq r_{il}$. This means that if j and k are between i and l, then j and k are at least as similar as i and l.

In general (Hubert (1974)), the goal of seriation in this context is to find an ordering or permutation of the objects being sequenced so that if this permutation is applied simultaneously to both rows and columns of the similarity matrix R, the resulting matrix is in strong Robinson form. The ordering is thought of as the natural order determined by the similarity data. The obvious question to ask is: when does there exist such a permutation and how does one find it? This question and this general approach to seriation is based on the work of Kendall (1963), (1969a,b), (1971a,b,c).

A square symmetric matrix is in *weak Robinson form* if the entries in any row do not decrease as the main diagonal is approached. This concept is due to Kendall (1969a) and is named after W. S. Robinson who used essentially this idea in sequence dating in archaeology (Robinson (1951)). It is easier to check whether a matrix has the weak Robinson form than it is to check for the strong Robinson form. Fortunately, it is not hard to show that for square symmetric matrices, the weak and strong notions of Robinson form coincide. For example, weak implies strong since $i \leqq j \leqq k \leqq l$ implies $r_{jk} \geqq r_{ik} \geqq r_{il}$. Henceforth, we shall use the term *Robinson form* for both the weak and strong notions.

We shall ask when there is a permutation putting a matrix in Robinson form. To give an example, let

$$R=\begin{pmatrix} 10 & 9 & 8 \\ 9 & 10 & 7 \\ 8 & 7 & 10 \end{pmatrix}. \tag{4.5}$$

Then R does not have the Robinson form, because $1 \leqq 2 \leqq 3 \leqq 3$, but $r_{23} < r_{13}$. This can also be seen because the entries in the third row decrease as the diagonal is approached. However, if we switch the first two rows and first two columns, we obtain the following matrix, which is in Robinson form:

$$\begin{pmatrix} 10 & 9 & 7 \\ 9 & 10 & 8 \\ 7 & 8 & 10 \end{pmatrix}.$$

Let ε be an arbitrary positive number, representing a threshold. Define a graph G_ε from R as follows. The vertices of G_ε are the numbers $1, 2, \cdots, n$, where n is the number of rows of R. There is an edge from i to j if and only if $r_{ij} \geqq \varepsilon$. If R is in Robinson form, then it is easy to see that G_ε satisfies the following condition: whenever $i \leqq j \leqq k \leqq l$ and there is an edge in G_ε between i

and l, then there is an edge in G_ε between j and k. Conversely, if we can verify this condition for every G_ε, then it follows that R is in Robinson form.

Speaking more generally, suppose G is an arbitrary graph and $\lesssim$ is an ordering of the vertices of G. We say that $\lesssim$ is *compatible* with G if whenever $i \lesssim j \lesssim k \lesssim l$ and there is an edge in G between i and l, then there is an edge in G between j and k. For example, in the graph of Fig. 4.8., if we assume a loop at each vertex, then a compatible ordering of the vertices is given by c, b, e, a, d. The ordering c, b, a, e, d is not compatible since $b \lesssim b \lesssim a \lesssim e$ and there is an edge between b and e, but no edge between b and a.

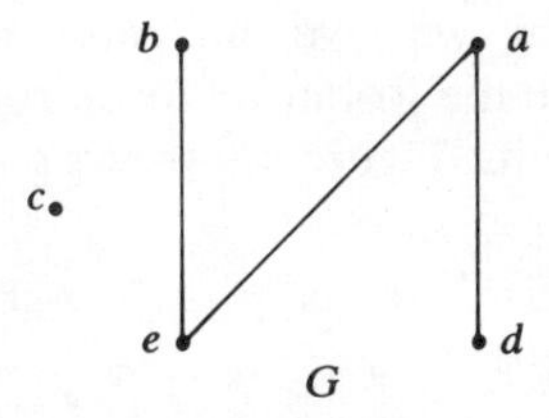

FIG. 4.8. *A compatible ordering for the graph G is given by c, b, e, a, d.*

THEOREM 4.3 (Roberts (1971b)). *A graph G (with a loop at each vertex) has some compatible ordering of the vertices if and only if G is an indifference graph.*

We shall use this result to get a complete characterization of the matrices permutable into Robinson form. The basic ideas behind this argument are contained in Hubert (1974). Let R' be obtained from R by replacing each diagonal element by ∞, or by a sufficiently high number. Then clearly R is permutable to Robinson form if and only if the diagonal elements of R are each maximal in their row and R' is permutable to Robinson form. Let G'_ε be defined from R' as was G_ε from R. Then clearly each G'_ε has a loop at each vertex. Hence, by Theorem 4.3, G'_ε has a compatible vertex ordering if and only if G'_ε is an indifference graph. We notice that if R' is permutable to Robinson form, then the same ordering of vertices is compatible with each G'_ε. In such a case, we say that the family of indifference graphs $\{G'_\varepsilon\}$ is *homogeneous.* (A similar idea arises in the study of probabilistic consistency—see Roberts (1971a).)

THEOREM 4.4. *R is permutable to Robinson form if and only if the diagonal elements of R are each maximal in their row and $\{G'_\varepsilon\}$ is a homogeneous family of indifference graphs.*

Notice that there are really only finitely many different graphs in the family $\{G'_\varepsilon\}$, so the criterion in the theorem is usable.

It is interesting to notice the connection between the ideas of this section and the notion of the consecutive 1's property defined in § 3.4. If G is a graph with n vertices, its *adjacency matrix* is an $n \times n$ matrix whose i, j entry is 1 if there is an edge between vertices i and j, and 0 if there is no such edge.

THEOREM 4.5 (Roberts (1968)). *Suppose G is a graph (with a loop at each vertex) and A is its adjacency matrix. Then G is an indifference graph if and only if A has the consecutive 1's property.*

THEOREM 4.6 (Kendall (1969a)). *Suppose K is any matrix of* 0's *and* 1's *which has the consecutive* 1's *property. Then the permutations of the rows of K which produce consecutive* 1's *correspond exactly to those permutations which when applied simultaneously to both rows and columns put KK^T into Robinson form.*

COROLLARY 1. *If G is an interval graph and M is its dominant clique-vertex incidence matrix, then MM^T is permutable to Robinson form.*

Proof. This follows by Theorem 3.4.

COROLLARY 2. *If G is an indifference graph and A is its adjacency matrix, then AA^T is permutable to Robinson form.*

4.3. Trees. In this section, we give one more necessary condition for the existence of a permutation to the Robinson form. A graph G is called a *tree* if G is connected but has no circuits. Figure 4.9 shows a variety of trees.

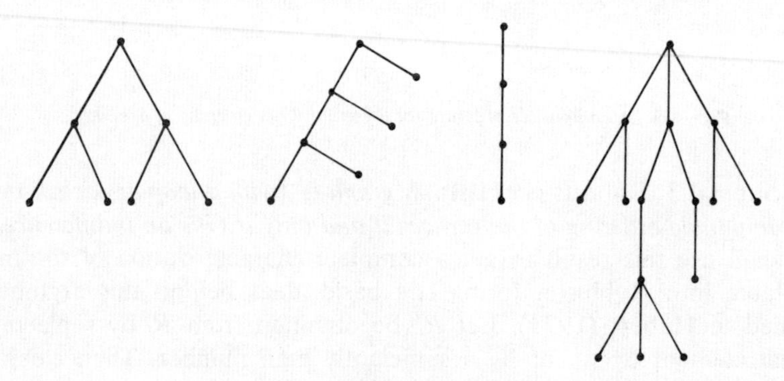

FIG. 4.9. *A variety of trees.*

THEOREM 4.7. *A graph is connected if and only if it has a spanning subgraph which is a tree.*

Proof. Clearly every graph which has a spanning subgraph which is a tree—we shall refer to a *spanning tree*—is connected. Conversely, suppose G is connected. Since the graph consisting of one vertex is a tree, certainly G has a subgraph which is a tree. Let H be a subgraph which is a tree and which has as many vertices as possible. If there are vertices in G but not in H, then by connectedness of G there must be an edge from some one of these vertices u to a vertex v of H. Adding the vertex u and the edge $\{u, v\}$ gives rise to a subgraph of G which is a tree and has more vertices than H, which is a contradiction. Hence, we conclude that every vertex of G is in H and H is spanning. Q.E.D.

Remark. In the depth first search procedure described in § 2.3, the edges used in labeling define a spanning tree.

Suppose we place a weight or real number on each edge of a connected graph. It is often important to find a *maximal* (or *minimal*) *spanning tree*, i.e., a spanning tree so that the sum of the weights of its edges is as large (as small) as possible. A large number of operations research problems boil down to the determination of maximal (or minimal) spanning trees. One algorithm for finding a maximal spanning tree is the *greedy algorithm*: initially pick any edge of largest weight; once $k-1$ edges have been chosen forming a tree T_{k-1}, add an edge of largest weight between a vertex of T_{k-1} and a vertex not in T_{k-1}. This forms the tree T_k. Once each vertex has been included, we have a maximal spanning tree. For example, in the graph of Fig. 4.10 (ignoring loops), we would first choose edge $\{2, 3\}$, then edge $\{3, 4\}$, and finally edge $\{1, 2\}$ to obtain a maximal spanning tree.

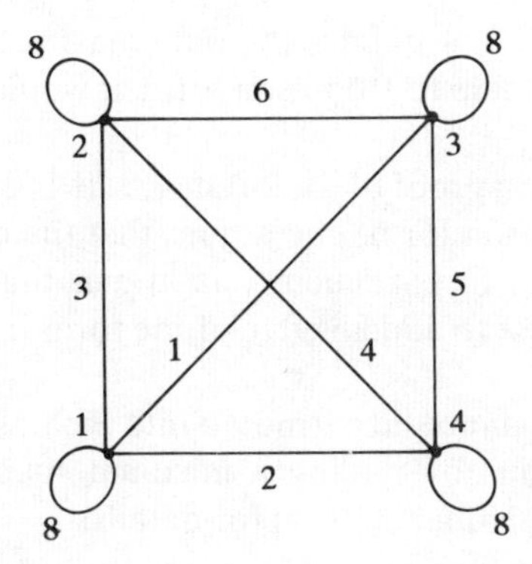

FIG. 4.10. *A maximal spanning tree is given by edges* $\{2, 3\}$, $\{3, 4\}$, *and* $\{1, 2\}$.

Suppose R is a square symmetric matrix of n rows. Let $G(R)$ be a graph with vertices $1, 2, \cdots, n$, an edge between i and j if and only if $r_{ij} \neq 0$, and a weight of r_{ij} on the edge $\{i, j\}$.

THEOREM 4.8 (Wilkinson (1971)). *If R is permutable to Robinson form, then some maximal spanning tree of $G(R)$ is a chain and the ordering of rows and columns putting R into Robinson form corresponds to the ordering of vertices in a maximal spanning tree which is a chain.*

To illustrate this theorem, we show in Fig. 4.11 the graph $G(R)$ obtained from the matrix R of (4.5). The permutation 2, 1, 3 which puts R into Robinson form corresponds to a maximal spanning tree consisting of edges $\{1, 2\}$ and $\{1, 3\}$. The

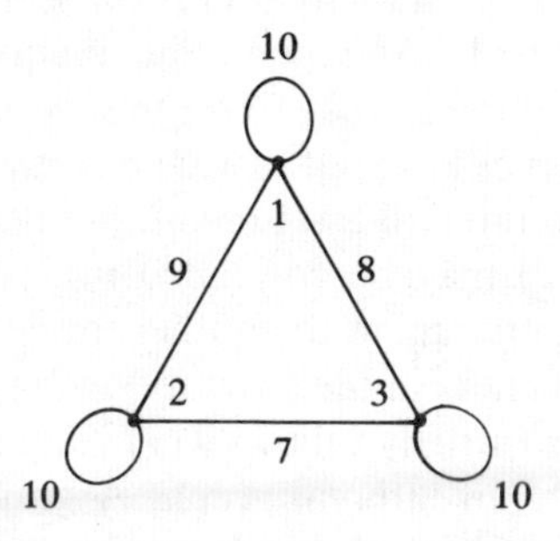

FIG. 4.11. *The graph $G(R)$ corresponding to the matrix R of* (4.5).

converse of Wilkinson's theorem is false. Suppose the graph of Fig. 4.10 is thought of as the graph $G(R)$ for a matrix R. Then R is the following matrix:

$$R = \begin{pmatrix} 8 & 3 & 1 & 2 \\ 3 & 8 & 6 & 4 \\ 1 & 6 & 8 & 5 \\ 2 & 4 & 5 & 8 \end{pmatrix}.$$

Note that the chain 1, 2, 3, 4 defines the only maximal spanning tree in $G(R)$. Hence, if R were permutable to Robinson form, it would already be in Robinson form. However, $r_{14} > r_{13}$.

It is not hard to supply a proof of Theorem 4.8. If $1, 2, \cdots, n$ is a permutation which puts R into Robinson form, one argues that the chain $1, 2, \cdots, n$ defines a maximal spanning tree. In particular, one argues that in applying the greedy algorithm, it is always best to add an edge of the form $\{i, i+1\}$. Details are left to the reader.

In practice, even if R cannot be permuted into Robinson form, one tries to get "close." One often begins by finding a maximal spanning tree in $G(R)$ and working from there. See Hubert (1974) for details.

4.4. Uniqueness. In seriation, it is possible to end up with more than one proper serial order. For example, a graph may have more than one compatible vertex ordering. Or, if we perform seriation using interval assignments as described in § 3.3, there may be two essentially different interval assignments. Here, we shall comment briefly on the uniqueness of these seriations.

If $1, 2, \cdots, n$ is a compatible vertex ordering for a graph, then $n, n-1, \cdots, 1$ is another such ordering. Thus, every graph of more than one vertex always has at least two compatible orderings. It is easy to show that if two vertices x and y are equivalent in the sense of the equivalence relation defined in § 4.1, then their places in the ordering may be interchanged. Moreover, vertices in a connected component must appear together, but the order in which components appear may change. However, up to complete reversal of the ordering, these are the only kinds of changes possible.

THEOREM 4.9 (Roberts (1971b)). *An indifference graph G (with a loop at each vertex) having more than two vertices has exactly two compatible vertex orderings (one the reverse of the other) if and only if G is connected and reduced.*

In the case of an interval assignment, the uniqueness question is a little trickier. Two interval assignments may differ in several ways. First, on nonoverlapping intervals, the interval assigned to x, $J(x)$, may *strictly follow* the interval assigned to y, $J(y)$, in one interval assignment, but not in the other. Second, on overlapping intervals, $J(x)$ may be contained in $J(y)$ in one assignment, but simply overlap without containment in another, or contain $J(y)$ in still another, etc. The question of uniqueness in the strict following sense may be looked at as follows. The relation "$J(x)$ strictly follows $J(y)$" defines a transitive orientation on the complement G^c of an interval graph. A graph with at least one edge which has a transitive orientation always has at least two: simply reverse all the directions in the orientation. Every transitive orientation of G^c gives rise to an interval assignment if G is an interval graph, through the procedure outlined in § 3.4. Hence, every interval graph with an edge always has at least two interval assignments which differ on strict following. Golumbic (1977) has recently found methods for identifying the number of different transitive orientations of a graph, and these methods allow the computation of the number of interval assignments which differ on strict following. Shevrin and Filippov (1970) and Trotter et al. (1976) have independently given a criterion for a graph to have exactly two transitive orientations, one obtained from the other by reversing directions. We shall present their result. If G is a disconnected, transitively orientable graph in which at least two different components each have an edge, then there are at least two distinct transitive orientations of each component with an edge. Two such orientations can be combined arbitrarily, producing at least four transitive orientations for G. Thus, we may restrict ourselves to connected transitively orientable graphs. A set K of vertices of a graph is called *partitive* if for every x, y in K and every u not in K, there is an edge from x to u if and only if there is an edge from y to u. A set of vertices is *independent* if there are no edges joining any of the vertices in the set.

THEOREM 4.10 (Shevrin and Filippov, Trotter, Moore, and Sumner). *A connected transitively orientable graph G with n vertices and at least one edge has exactly two transitive orientations (one obtained from the other by reversing directions) if and only if every partitive set K with at least two vertices but fewer than n vertices is independent.*

To illustrate this theorem, note that the complete graph on 3 vertices has three transitive orientations. Any two vertices form a partitive set which is not independent.

Two interval assignments can differ in more than just strict following. The problem of uniqueness of interval assignments still requires a treatment of the uniqueness of the relations among overlapping intervals. For example, when is $J(x)$ *always* contained in $J(y)$ in every interval assignment? This problem has not yet been worked out.

CHAPTER 5

Food Webs, Niche Overlap Graphs, and the Boxicity of Ecological Phase Space

5.1. Boxicity. A very interesting class of intersection graphs is that class arising from the boxes in Euclidean n-space. A *box* is just a generalized rectangle with sides parallel to the coordinate axes, for example a rectangle in 2-space. We shall show below that every graph G is the intersection graph of boxes in some n-space. The smallest such n is called the *boxicity* of G, and is abbreviated *box* (G). It is convenient to think of complete graphs as being intersection graphs of boxes in 0-space (each box here is a single point[12]), and hence having boxicity 0. The interval graphs are exactly the graphs of boxicity $\leqq 1$.

Let us calculate the boxicity of various graphs. The graph Z_4 is not an interval graph, and hence has boxicity larger than 1. It is easy to show that Z_4 is the intersection graph of boxes in 2-space (see Fig. 5.1) and hence that box $(Z_4) = 2$. Indeed, box $(Z_n) = 2$, all $n \geqq 4$.

Another class of graphs for which it is not hard to calculate the boxicity is the class of complete p-partite graphs. A graph G is *complete p-partite* if the vertices are partitioned into classes $N_1, N_2, \cdots, N_p$ and for every i, j there is an edge joining vertex x of N_i and vertex y of N_j iff $i \neq j$. If N_i has n_i vertices, we denote the corresponding graph $K(n_1, n_2, \cdots, n_p)$. The graph $K(2, 2)$ is really Z_4, with the vertices a and c (as labeled in Fig. 5.1) forming one class and the remaining vertices the other class. The graph $K(3, 3)$ is the famous water-light-gas graph: there are three houses and three utilities, and each house is hooked up to each utility. The graph $K(1, 1, \cdots, 1)$ with p classes is the complete graph with p vertices, which is usually denoted K_p. It is easy to see that box $K(n_1, n_2, \cdots, n_p) \leqq p$. A construction for $K(3, 3)$ is shown in Fig. 5.2, and it is easy to see how this generalizes. In general, box $K(n_1, n_2, \cdots, n_p, 1) =$ box $K(n_1, n_2, \cdots, n_p)$. For, having obtained a representation for $K(n_1, n_2, \cdots, n_p)$ with boxes in n-space, one obtains one for $K(n_1, n_2, \cdots, n_p, 1)$ with boxes in n-space by adding one huge box which contains all of the others.

THEOREM 5.1 (Roberts (1969b)). *If $G = K(n_1, n_2, \cdots, n_p)$, then* box $(G) =$ *the number of n_i which are greater than* 1.

As a special case, we note that since the complete graph K_p is $K(1, 1, \cdots, 1)$, with p classes, Theorem 5.1 gives us box $(K_p) = 0$.

In general, it is hard to compute the boxicity of a graph. It is an unsolved problem to find a procedure for calculating boxicity. It is also an unsolved

[12] We do not require the boxes corresponding to distinct vertices to be distinct.

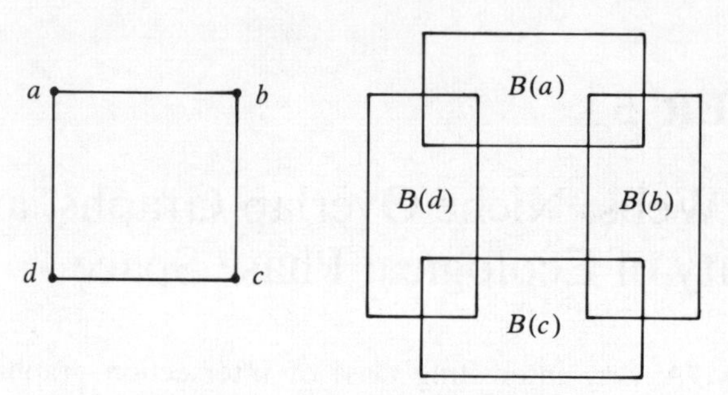

FIG. 5.1. *The boxicity of Z_4 is 2.*

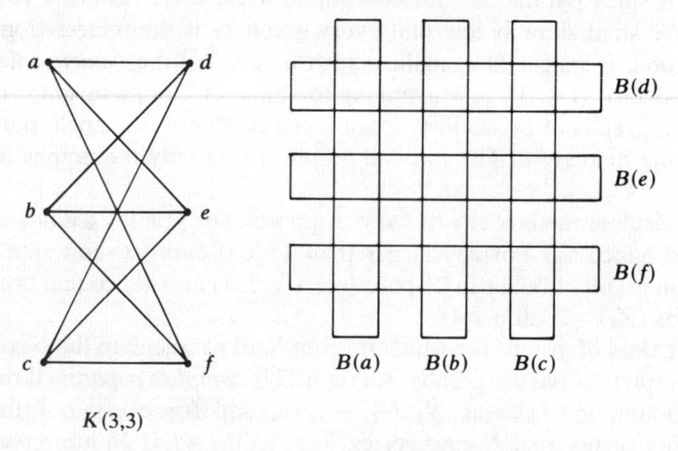

FIG. 5.2. *The boxicity of $K(3, 3)$ is at most 2.*

problem to characterize the graphs of given boxicity, for example 2. We shall see below why these two problems are important. Recently, Gabai (1974) has obtained some results on boxicity. To present them, we say that a set of edges in a graph is *independent* if no two edges in the set have a vertex in common.

THEOREM 5.2 (Gabai (1974)). *If the maximum cardinality of a set of independent edges of G^c is m, then* $\text{box}(G) \leqq m$. *Moreover, if G^c has a generated subgraph of k independent edges, then* $\text{box}(G) > k-1$.

The second statement in Theorem 5.2 simply says that if G has $K(2, 2, \cdots, 2)$ with k classes as a generated subgraph, then $\text{box}(G) > k-1$. This follows immediately from Theorem 5.1 since the boxicity of a graph is at least as great as the boxicity of any generated subgraph. The first part of the theorem shows that $\text{box}(Z_4) \leqq 2$. For, using the labeling of vertices in Fig. 5.1, we note that the maximum independent set of edges in G^c consists of the two independent edges $\{a, c\}$ and $\{b, d\}$. The second part of Gabai's result shows that $\text{box}(Z_4) > 1$, for Z_4^c is a graph consisting of 2 independent edges. Hence, $\text{box}(Z_4) = 2$, as we have already observed.

THEOREM 5.3 (Roberts (1969b)). *Every graph G of n vertices is the intersection graph of boxes in n-space.*

Proof. Note that every box in n-space is defined by giving n intervals, $J_1, J_2, \cdots, J_n$, the projections of the box onto the different coordinate axes. In particular, if B' is another box with corresponding intervals $J'_1, J'_2, \cdots, J'_n$, then

$$B \cap B' \neq \varnothing \quad \text{iff} \quad (\forall i)(J_i \cap J'_i \neq \varnothing).$$

Let $u_1, u_2, \cdots, u_n$ be the vertices of G, and for every i and k, let

$$J_i(u_k) = \begin{cases} [0, 1] & \text{if } k = i, \\ [1, 2] & \text{if } k \neq i \text{ and } \{u_i, u_k\} \in E(G), \\ [2, 3] & \text{if } k \neq i \text{ and } \{u_i, u_k\} \notin E(G). \end{cases}$$

Note that for all $j \neq k$,

$$\{u_j, u_k\} \in E(G) \quad \text{iff} \quad (\forall i)(J_i(u_j) \cap J_i(u_k) \neq \varnothing).$$

Thus, we may take the intervals $J_1(u_k), J_2(u_k), \cdots, J_n(u_k)$ to define a box $B(u_k)$ in n-space, and we have for all $j \neq k$,

$$\{u_j, u_k\} \in E(G) \quad \text{iff} \quad B(u_j) \cap B(u_k) \neq \varnothing. \quad \text{Q.E.D.}$$

5.2. The boxicity of ecological phase space. Ecologists study the relationships among organisms in communities. From the point of view of biology, this study is a crucial step in the understanding of the "web of nature." From the point of view of society, this study is important to understand how social endeavors would perturb ecosystems. In ecology, a species[13] is sometimes characterized by the ranges of all of the different environmental factors which define its normal healthy environment. For example, the normal healthy environment is determined by a range of values of temperature, of light, of pH, of moisture, and so on. If there are n factors in all, and each defines an interval of values, then the corresponding region in n-space is a box. This box corresponds to what is frequently called in ecology the *ecological niche* of the species. Hutchinson (1944), for example, defines the ecological niche as "the sum of all the environmental factors acting on an organism; the niche thus defined is a region of n-dimensional hyper-space, comparable to the phase-space of statistical mechanics." For this reason, the n-dimensional Euclidean space defined by the n factors is sometimes called *ecological phase space.* Recent reviews of the concept of ecological niche are by Miller (1967), Vandermeer (1972), and Pianka (1976).

Suppose we have some independent information about when different species' niches overlap. We can then ask how many dimensions are required of an ecological phase space so that we can represent each species by a niche or box in this space and so that the niches overlap if and only if the independent information tells us they should. This question can be formulated graph-theoretically. Draw a *niche overlap graph* whose vertices are a collection of species from an ecosystem, and which has an edge between two species if and only if their ecological niches overlap. We wish to determine the smallest n so that the niche

[13] We use this term loosely.

overlap graph is the intersection graph of boxes in n-space, i.e., we wish to determine the boxicity of the niche overlap graph.

How do we define the niche overlap graph? One way is to use the notion of competition. For it is an old ecological principle that two species compete if and only if their ecological niches overlap. By studying the species in an ecosystem, we can obtain information about which species prey on which others. We can represent this information in a digraph, whose vertices are the species in question and which has an arc from species x to species y if and only if x preys on y. This digraph is called a *food web.* Given a food web, we shall assume, following Cohen (1978), that two species have overlapping niches (at least along trophic or "feeding" dimensions) if and only if they have a common prey. That is, niche overlap occurs if and only if the species compete for food. Figures 5.3 and 5.4 show two food webs and their corresponding niche overlap graphs. Note, for example, that in Fig. 5.3, species 1 and 4 both prey on species 5; hence there is an edge between vertices 1 and 4 in the niche overlap graph.[14]

The niche overlap graph of Fig. 5.3 is a rather simple graph, and it is easy to see that it is an interval graph which is incomplete. Hence, its boxicity is 1. The conclusion is that one dimension suffices to account for niche overlap in this ecosystem. However, the nature of this dimension, i.e., its environmental significance, is not determined. The niche overlap graph of Fig. 5.4 is a little more complicated. However, surprisingly, it too is an interval graph. Figure 5.5 shows an interval assignment. Hence, the boxicity is again 1.

The idea of studying the dimensionality or boxicity of the ecological phase space needed to account for observed niche overlaps was developed by Joel Cohen in an unpublished document in 1968. After examining a number of real food webs, Cohen discovered that all of them gave rise to niche overlap graphs which were interval graphs. He asked whether this was always true, and Klee (1969) published the question. Should every niche overlap graph be an interval graph, then one dimension would suffice to account for niche overlap, a very surprising development. This dimension, though not defined, might have great ecological significance, and might help in understanding perturbations in ecosystems. Until recently, the evidence mounted in favor of the conjecture that every niche overlap graph obtained from a real food web is an interval graph. However, in Cohen (1978), a number of counterexamples are finally given. In that monograph, a large number of food webs are examined, and the evidence seems to suggest that food webs corresponding to habitats of a certain limited physical and temporal heterogeneity do give rise to niche overlap graphs which are interval graphs. Cohen also develops statistical profiles of food webs, with the aim of generating hypothetical food webs at random in order to generate their corresponding niche overlap graphs. These graphs can then be used to study various claims about dimensionality of niche overlap. In order to study

---

[14] In his early work, Cohen identified niche overlap with competition and called the niche overlap graph the *competition graph.* In Cohen (1978), he argues that competition may occur for resources other than food and hence the term niche overlap graph is more appropriate.

niche overlap further, it would be nice to have techniques for calculating boxicity. Unfortunately, as pointed out above, such techniques are lacking.

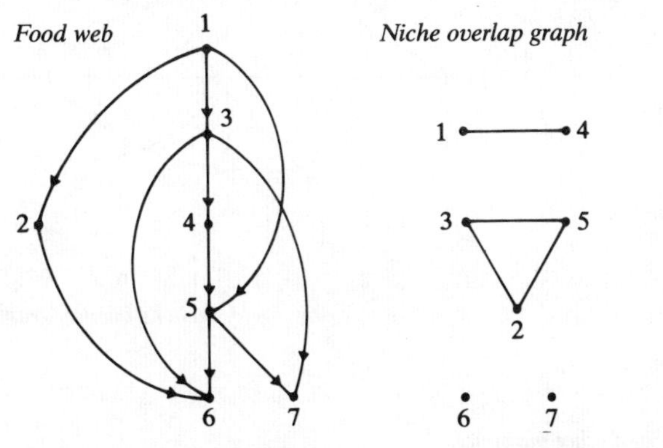

FIG. 5.3. *Food web and niche overlap graph for the Strait of Georgia, British Columbia. From data of Parsons and LeBrasseur* (1970), *as adapted by Cohen* (1978).

5.3. The properties of niche overlap graphs. It seems natural to ask whether we can explain structurally the empirical results described in the previous section. Is it so surprising that so many real niche overlap graphs are interval graphs, or are almost all possible niche overlap graphs interval graphs? To make this question precise, let us assume that a food web F corresponds to an *acyclic digraph*, i.e., a digraph with no cycles. This is a special assumption, but one which is commonly made in ecology. The corresponding niche overlap graph G is defined as follows: the vertices of G are the vertices of F and there is an edge from species x to species y if and only if for some z, there are arcs (x, z) and (y, z) in F. Given an arbitrary graph G, we say it is a *niche overlap graph* if it comes from an acyclic digraph (or food web) in this way. What graphs are niche overlap graphs? Are almost all niche overlap graphs interval graphs?

To answer these questions, let us first observe that not every graph is a niche overlap graph. For, it is easy to see that every acyclic digraph F must have a vertex with no outgoing arcs. This vertex corresponds to an *isolated vertex* in the niche overlap graph of F, i.e., to a vertex with no adjacent vertices. Hence, every niche overlap graph has an isolated vertex. Let I_p be the graph of p isolated vertices. We next observe that if G is any graph at all, and G has e edges, then $G \cup I_e$ is a niche overlap graph, where $G \cup I_e$ is the graph obtained by adding e isolated vertices to G. To see this, we build a food web F as follows. We start by putting into F the vertices of G. For every edge $\alpha = \{a, b\}$ of G, we add a vertex x_α to F. In F, we include arcs from a and b to x_α. It is clear that $G \cup I_e$ is the niche overlap graph of F. This result shows that essentially every graph is a niche overlap graph. In particular, there are niche overlap graphs of arbitrarily high boxicity. For example, take $K(2, 2, \cdots, 2) \cup I_e$. Hence, it is surprising, given that almost every possible graph is potentially a niche overlap graph, that

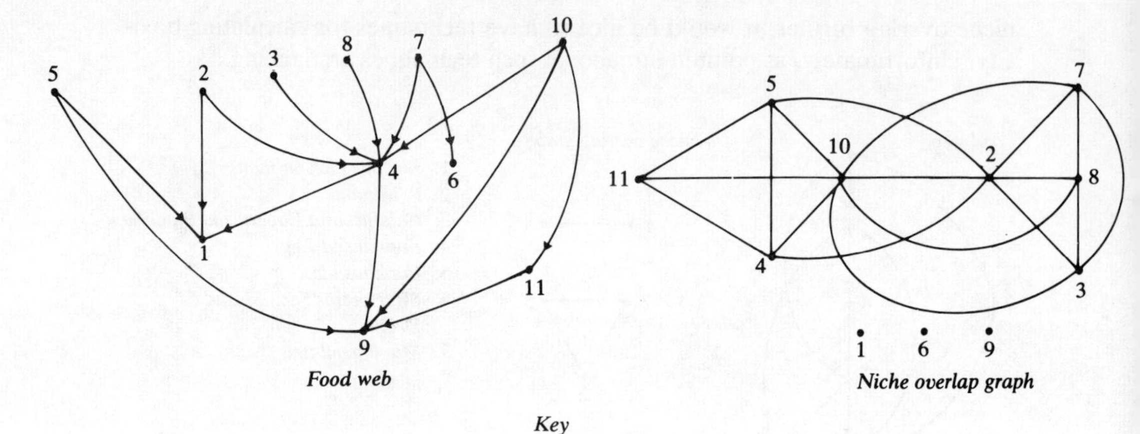

Key

1. *Canopy—leaves, fruits, flowers*
2. *Canopy animals—birds, fruit-bats, and other mammals*
3. *Upper air animals—birds and bats, insectivorous*
4. *Insects*
5. *Large ground animals—large mammals and birds*
6. *Trunk, fruit, flowers*
7. *Middle-zone scansorial animals—mammals in both canopy and ground zones*
8. *Middle-zone flying animals—birds and insectivorous bats*
9. *Ground—roots, fallen fruit, leaves and trunks*
10. *Small ground animals—birds and small mammals*
11. *Fungi*

FIG. 5.4. *Food web and niche overlap graph for Malaysian Rain Forest, from data of Harrison* (1962), *as adapted by Cohen* (1978).

real-world niche overlap graphs tend to have such low boxicity, and usually boxicity 1.

The *niche overlap number* (or *competition number*) $k(G)$ is the least number k so that $G \cup I_k$ is a competition graph. Characterization of niche overlap graphs is equivalent to the problem of computing $k(G)$. We shall present some results on $k(G)$.

For the rest of this section $e(G)$ will denote the number of edges of G and $n(G)$ the number of vertices.

THEOREM 5.4 (Roberts (1978)). *If G has no triangles, then* $k(G) \geqq e(G) - n(G) + 2$.

Proof. Let $n = n(G)$, $e = e(G)$, and $k = k(G)$. Suppose $G \cup I_k$ is a niche overlap graph for food web F. For every edge $\alpha = \{u, v\}$ of G, there is a vertex a_α

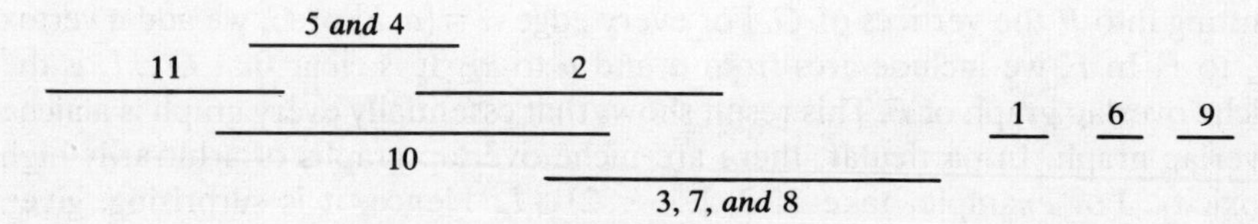

FIG. 5.5. *Interval assignment for the niche overlap graph of Fig. 5.4.*

in F so that u and v prey on a_α. Moreover, since G has no triangles, the a_α's are distinct. Hence, it follows that the number of vertices of F, namely $n+k$, is at least e. Now every a_α has at least two incoming arcs, and it is not hard to show that every acyclic digraph has at least two vertices with fewer than two incoming arcs. For, by a variant of an earlier observation, every such digraph D has a vertex x with no incoming arcs. The digraph D' obtained by removing x and all outgoing arcs from x is acyclic and hence has a vertex y with no incoming arcs. Hence, x and y have fewer than 2 incoming arcs in D. In any case, these two vertices cannot be used for a_α's, and so we conclude that $n+k-2 \geqq e$. Q.E.D.

The bound in Theorem 5.4 is not sharp. The graph I_3 has no triangles, yet $k=0$ while $e-n+2=-1$. However, we have the following theorem.

THEOREM 5.5 (Roberts (1978)). *If G is connected, $n(G)>1$, and G has no triangles, then $k(G)=e(G)-n(G)+2$.*

We shall give a hint of the method of proof of this theorem below. Note that by Theorem 5.5, $k(Z_n)=2$ if $n \geqq 4$, since $e=n$.

THEOREM 5.6 (Roberts (1978)). *For every k, there is a graph G with $k(G)>k$.*

Proof. Let $k \geqq 4$, and let G consist of two copies of Z_k with vertices in one copy joined to corresponding vertices in the other copy. Then G has no triangles, $e(G)=3k$, $n(G)=2k$, and so $k(G) \geqq 3k-2k+2=k+2$. Q.E.D.

If a is a vertex of a graph, the *open neighborhood of a*, $N(a)$, is the collection of all vertices adjacent to a. The vertex is called *simplicial* if $N(a)$ is a clique. (The extreme vertices of a graph which we encountered in § 4.1 are simplicial vertices.)

THEOREM 5.7 (Dirac (1961)). *Every rigid circuit graph has a simplicial vertex.*

THEOREM 5.8 (Roberts (1978)). *If G is a rigid circuit graph, then $k(G) \leqq 1$.*

Proof. Find a sequence of vertices $a_1, a_2, \cdots, a_{n-1}$, where $n=n(G)$, as follows. Vertex a_1 is a simplicial vertex of $G_1=G$. Vertex a_2 is a simplicial vertex of $G_2=G_1-a_1$, the graph generated by vertices of G_1 other than a_1. (Note that a_2 exists since every generated subgraph of a rigid circuit graph is rigid circuit.) Vertex a_3 is a simplicial vertex of $G_3=G_2-a_2$. And so on. Build a sequence of food webs $F_1, F_2, \cdots, F_{n-1}$ as follows. Food web F_i for all i has vertex set $V(G)$ plus one additional vertex x. Food web F_1 has arcs from a_1 and all vertices in $N(a_1)$ to x. Thus, a_1 and all vertices in $N(a_1)$ are all adjacent to each other in the niche overlap graph corresponding to F_1. F_2 is obtained from F_1 by adding arcs from a_2 and all vertices of $N(a_2)$ to a_1. Thus, the niche overlap graph of F_2 is obtained from that of F_1 by joining a_2 and all its neighbors to each other. In general, F_{i+1} is obtained from F_i by adding arcs from a_{i+1} and all vertices of $N(a_{i+1})$ to a_i. It is now easy to see that food web F_{n-1} is acyclic and its niche overlap graph is $G \cup I_1$, where vertex x corresponds to the isolated vertex. Q.E.D.

The procedure of proof of this theorem is carried out on an example in Fig. 5.6.

COROLLARY. *If G is an interval graph, then $k(G) \leqq 1$.*

The construction of the proof of Theorem 5.8 depends on an ordering of vertices. For each possible ordering of vertices of *any* graph, we can carry out a similar construction, if we are careful to divide up $N(a_i)$ into cliques, and if we allow ourselves to add more than one additional isolated vertex. In this way, one obtains an estimate for $k(G)$. This procedure leads to the result of Theorem 5.5 for connected, triangle-free graphs, and to estimates of $k(G)$ for other graphs. The reader is referred to Roberts (1978) for details.

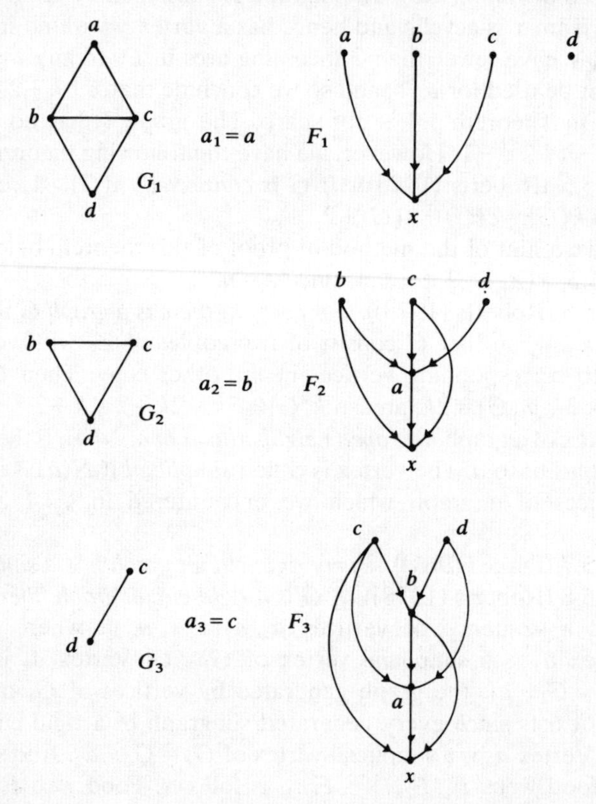

FIG. 5.6. *Constructing a food web for $G \cup I_1$ if G is a rigid circuit graph.*

5.4. Community food webs, sink food webs, and source food webs. Certain food webs include the predation relations between all pairs of species. These food webs Cohen (1978) calls *community food webs.* Suppose F is a community food web, and W is a set of species from F. Let X be the set of all species which are reachable from vertices of W by a path, and Y the set of all species which reach vertices of W by a path. The subgraph (subdigraph) of F generated by vertices of X is called a *sink food web corresponding to* W and the subgraph generated by vertices of Y is called a *source food web.* The food web of Fig. 5.3 is a sink food web obtained from some larger food web, while that of Fig. 5.4 is a community food web. In the latter food web, if $W = \{1, 4, 10\}$, then $X = \{1, 4, 9, 10, 11\}$[15] and $Y = \{1, 2, 3, 4, 5, 7, 8, 10\}$.

[15] Every vertex x is reachable from itself since x alone is a path.

THEOREM 5.9 (Cohen (1978)). *A community food web has a niche overlap graph which is an interval graph if and only if every sink food web contained in it does.*

Proof. Every community food web is a sink food web: take $W = V$. Thus, it is only necessary to prove that if F is a community food web and its niche overlap graph G is an interval graph, then for every set W of vertices of F, if H is the niche overlap graph of the sink food web corresponding to W, then H is an interval graph. Let x and y be vertices in the set X of vertices reachable from vertices of W. Then x and y have a common prey in F if and only if they have a common prey in the sink food web generated by W. It follows that H is a generated subgraph of G. The theorem follows since every generated subgraph of an interval graph is an interval graph. Q.E.D.

It is possible for a community food web to have a niche overlap graph which is an interval graph while some source food web contained in it has a niche overlap graph which is not an interval graph. To give an example, consider the (community) food web of Fig. 5.7. The corresponding niche overlap graph, also shown in the figure, is an interval graph. The source food web corresponding to the vertices x, y, z, f has the niche overlap graph shown in the figure. This graph is not an interval graph, since Z_4 is a generated subgraph. In general, source food webs cannot be expected to tell us much of interest about the dimensionality of the ecological phase space required to represent niche overlap. Hence, we must be careful, in gathering empirical data, to be sure not to use source food webs.

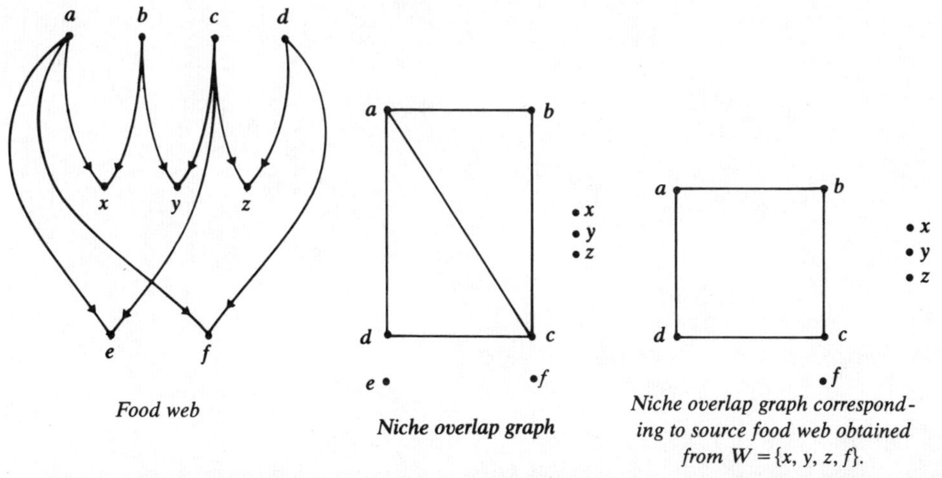

FIG. 5.7. *A community food web, its niche overlap graph which is an interval graph, and a niche overlap graph of a source food web which is not an interval graph.*

CHAPTER 6
Colorability

6.1. Applications of graph coloring. Suppose G is a graph. Let us try to color the vertices of G, each vertex receiving exactly one color, in such a way that if two vertices are joined by an edge, they get different colors. If such a vertex coloring can be carried out using k (or fewer) colors, we say that G is *k-colorable.* The smallest k so that G is k-colorable is called the *chromatic number* of G, and is denoted $\chi(G)$. It is obvious that $\chi(K_n)=n$ (each vertex must receive a different color). Moreover, if n is even, then $\chi(Z_n)=2$: alternate colors around the circuit. However, if n is odd, then $\chi(Z_n)=3$. (A coloring of Z_5 in 3 colors is shown in Fig. 6.1; no 2-coloring exists, as is easy to see.) Graph coloring problems arise in a variety of applications. We shall mention several of these here.

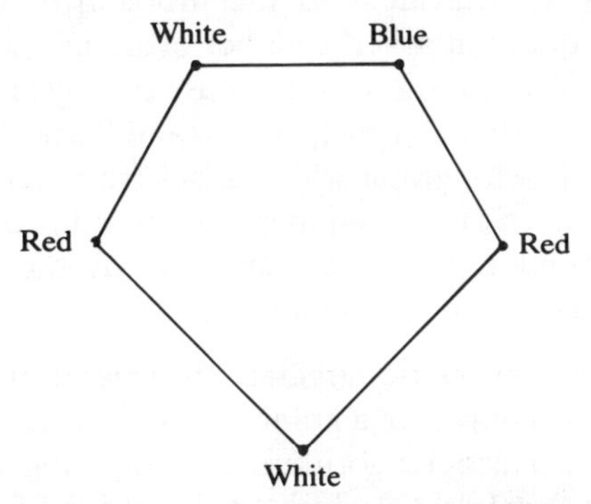

FIG. 6.1. *A coloring of Z_5 in 3 colors.*

6.1.1. Tour graphs. A *tour* of a garbage truck is a schedule of sites it visits in a given day. The following problem arose (Beltrami and Bodin (1973), Tucker (1973)) from a problem posed by the New York City Department of Sanitation. Given a collection of tours of garbage trucks, is it possible to assign each tour to a day of the week (other than Sunday) so that if two tours visit a common site, they get a different day? A similar problem can be posed for other service schedules, for example milk or newspaper deliveries, street cleaning schedules, and so on. To formulate this problem graph-theoretically, let G be the *tour graph,* the graph whose vertices are the tours, and which has an edge between two tours if and only if they visit a common site. The problem is equivalent to the following: is it possible to assign to each vertex (tour) one of six colors (days) so that if two tours are joined by an edge (visit a common site) they get a different color? Thus the question becomes: is the tour graph 6-colorable?

6.1.2. Committee schedules. Each member of one house of a state legislature belongs to several committees. A schedule of committee meetings is to be drawn up during a weekly period. Each committee is to meet exactly once, but two committees with a common member cannot meet at the same time. How many meeting times are required? To answer this question, we form a graph G with vertices the committees, and an edge between two committees if and only if their members overlap (this is the intersection graph of the committees, to use the terminology of § 3.2). We wish to assign to each vertex (committee) a color (meeting time) so that if two vertices are joined by an edge (have a member in common), they get different meeting times. The smallest number of meeting times required is the chromatic number of the graph G. A similar problem obviously arises in planning final examination schedules at a university. Here, the committees correspond to classes.

6.1.3. Map coloring. Given a map, we wish to use a variety of colors to color the countries, and we insist only that if two countries have a common boundary, they get different colors. We can translate a map into a graph by letting each country be represented by a vertex, and joining two vertices with an edge if and only if the corresponding countries have a common boundary. Then the problem of coloring the map is equivalent to the problem of coloring its graph. In particular, a famous question asked whether every map could be colored using four or fewer colors. This question, which was recently answered in the affirmative (Appel and Haken (1977), Appel, Haken and Koch (1977)), was equivalent to the question: could every graph which arises from a map be colored in four colors? (The graphs which arise from maps are called *planar*, as they are, under some reasonable assumptions about maps, exactly the graphs which can be drawn in the plane without edges crossing.)

6.2. Calculating the chromatic number. In general, it is a hard problem to calculate the chromatic number of a graph. In fact, it is not known whether there is a polynomial (deterministic) algorithm for computing $\chi(G)$. The problem of computing $\chi(G)$ is in the class *NP* which we defined in § 2.4. Stockmeyer (1973) shows that the problem of determining whether a planar graph is 3-colorable is *NP*-complete (and hence, of course, so is the problem of determining $\chi(G)$). Garey, Johnson and Stockmeyer (1976) have recently shown that even the problem of determining 3-colorability in planar graphs each of whose vertices has at most four adjacent vertices is *NP*-complete. These results show why such problems as routing of garbage trucks and scheduling of committee meetings are hard problems, for they correspond to coloring problems which are hard in a very precise way. It is interesting to note how translation of a problem into a precise mathematical one can cast light on why the problem is difficult.

It is easy to discover deterministically in polynomial time whether or not a graph is 2-colorable. For a graph is 2-colorable if and only if it is *bipartite*, i.e., the vertices can be partitioned into two classes, so that all edges in the graph go between classes. The depth-first search procedure described in § 2.3 can be used to find a polynomial deterministic algorithm for testing bipartiteness (Reingold, Nievergelt and Deo (1977, pp. 399, 400)).

The following theorem characterizes 2-colorable graphs.

THEOREM 6.1 (König (1936)). *A graph is 2-colorable if and only if it has no circuits of odd length.*

Proof. If G is 2-colorable, then every circuit must alternate between the two colors, and hence can only have even length. Conversely, suppose every circuit of G is even. We may assume without loss of generality that G is connected, for otherwise, we perform the 2-coloring separately on each component. If G is connected, we let $d(u, v)$ be the length of the shortest chain between u and v. Picking an arbitrary u in $V(G)$, we let

$$A = \{v \in V(G) : d(u, v) \text{ is even}\},$$

$$B = \{v \in V(G) : d(u, v) \text{ is odd}\}.$$

Notice that u is in A, since $d(u, u) = 0$. It is easy to show that there are no edges between vertices within class A or between vertices within class B. The argument proceeds by showing that if there were such an edge, then there would be an odd length closed chain in G. If there were an odd length closed chain, a shortest such would have to be an odd circuit. A similar argument shows that A and B are disjoint, and hence they can form the two classes of vertices in a 2-coloring. Q.E.D.

6.3. Clique number. The chromatic number is closely related to another number associated with a graph, the *clique number* $\omega(G)$, which is defined to be the size of the largest clique. The clique number arises in a number of applications. It is important in present day sociology, for example, where finding cliques in sociograms, graphs representing some relation among members in a group, is an important procedure. It is easy to see that $\chi(G) \geqq \omega(G)$, since every vertex of a clique must get a different color. It is also easy to see that χ may be larger than ω. For example, $\chi(Z_5) = 3$ while $\omega(Z_5) = 2$. A graph is called *weakly γ-perfect* if $\chi(G) = \omega(G)$. Thus, Z_5, or indeed Z_n, n odd and greater than 3, is not weakly γ-perfect. (These graphs Z_n are sometimes called *odd holes*.) The term γ-perfect arises from the fact that $\gamma(G)$ is sometimes used to denote the minimum number of independent sets which partition the vertices of G. It is easy to see that $\chi(G) = \gamma(G)$. The vertices of a given color form the independent sets.

If a graph is weakly γ-perfect, then of course its chromatic number can be calculated from its clique number. Although on the surface this seems like an easier problem, we have remarked earlier that the problem of finding the clique number is *NP*-hard. However, within the contexts of certain algorithms, it is easier to calculate clique number than chromatic number. In particular, in procedures for solving the tour graph problem described in § 6.1.1, determination of chromatic number must be done over and over for a continually changing set of tours. Since the set of tours is changed bit by bit, the tour graph gets changed only locally. This makes it easy to calculate clique number for subsequent graphs from previous ones by making only local searches, while it is not possible to calculate chromatic number of subsequent graphs by making only local searches. For this reason, Tucker (1973) suggests using clique number to calculate chromatic number. The procedure works only if there is a method of determining whether or not these two numbers are the same, i.e., whether or not a given graph is weakly γ-perfect. We shall discuss such a method.

We say that a graph is *γ-perfect* if every generated subgraph is weakly γ-perfect. The notion of γ-perfectness might seem rather restrictive, but we shall see that a large class of graphs are γ-perfect. The concept of γ-perfect is due to Berge (1961), (1962), who conjectured that a graph G is γ-perfect if and only if its complement is γ-perfect. This conjecture, known as the *weak Berge conjecture*, or *weak perfect graph conjecture*, was proved by Lovász.

THEOREM 6.2 (Lovász (1972a,b)). *A graph G is γ-perfect if and only if its complementary graph is γ-perfect.*

Since no odd hole is weakly γ-perfect, a γ-perfect graph cannot contain an odd hole as a generated subgraph, and neither, by Lovász' result, can its complementary graph. The converse of this statement was suggested by the work of Berge (1963), (1967), (1969), and is called the *strong Berge conjecture* or the *strong perfect graph conjecture.* It says that if neither G nor G^c contains an odd hole as a generated subgraph, then G is γ-perfect.

Tucker's (1973) idea is to use the strong Berge conjecture to determine whether or not $\chi(G)=\omega(G)$. If the answer is affirmative, then one can use the improved local procedure for finding $\omega(G)$ and use the result to calculate $\chi(G)$. As Tucker points out, the worst possible situation which could arise from the use of the Berge conjecture in sanitation scheduling is to obtain a set of tours which is supposedly assignable to the six days of the week, but which in fact cannot be so assigned. Then, one would have found a counterexample to the Berge conjecture! Note: use of the Berge conjecture involves determining the existence of odd holes. There can be exponentially many circuits in a graph (consider K_n, which has $\binom{n}{i}(i-1)!$ circuits of length i). However, we only need to identify circuits in a graph after local changes have been made in previous graphs.

6.4. γ-perfect graphs. Among the interesting classes of graphs which are γ-perfect are the bipartite graphs, the transitively orientable graphs, and the rigid circuit graphs (and hence the interval graphs and indifference graphs). We shall sketch a proof that every rigid circuit graph is γ-perfect. A *cutset* or an *articulation set* in a connected graph G is a set of vertices U such that the subgraph generated by $V(G)-U$ is disconnected. The vertices a and c form an articulation set in the graph of Fig. 6.2.

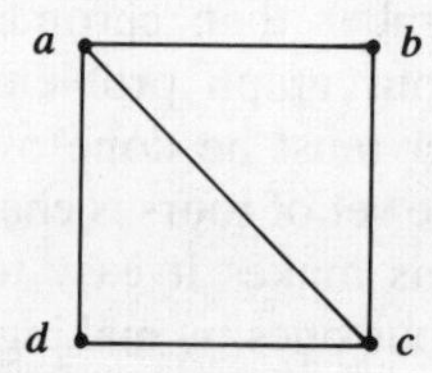

FIG. 6.2. *Vertices a and c form an articulation set.*

THEOREM 6.3 (Hajnal and Surányi (1958)). *In a connected rigid circuit graph G, every minimal articulation set is a clique.*

Proof. Let U be a minimal articulation set and H be the subgraph generated by $V(G)-U$. Then H has connected components $K_1, K_2, \cdots, K_p$, with $p \geqq 2$. Given u and v in U, we shall show they are joined by an edge in G. Now every vertex a in U has an edge to each K_i, otherwise $U-\{a\}$ would be an articulation set contained in U. Hence there must be x and y in K_1 so that $\{u, x\}$ and $\{v, y\}$ are edges of G. Since K_1 is connected, there is a chain $x_1, x_2, \cdots, x_r$ from x to y in K_1. Thus, there is a chain $u, x_1, x_2, \cdots, x_r, v$ with $x_1, x_2, \cdots, x_r$ in K_1. Let C be such a chain of minimal length. Let C' be a similar minimal length chain $u, y_1, y_2, \cdots, y_s, v$ with $y_1, y_2, \cdots, y_s$ in K_2. Then C followed by C' is a circuit in G. Since G is a rigid circuit graph, this cannot be a generated subgraph, and hence there must be some edge joining two vertices in this circuit. By minimality of C and C', there are no edges joining vertices on C except possibly u to v, and similarly for C'. Moreover, since K_1 and K_2 are different components of H, there are no edges joining any x_i to any y_j. Thus, the only possible edge in this circuit is the edge $\{u, v\}$. Q.E.D.

COROLLARY (Berge). *Every rigid circuit graph G is γ-perfect.*

Proof. It is sufficient to show that G is weakly γ-perfect, since every generated subgraph of a rigid circuit graph is also rigid circuit. It is also sufficient to assume that G is connected. The proof is by induction on the number of vertices of G. The case where G has one vertex is trivial. Assume the result is true for graphs with fewer vertices than G. If G is complete, the result is trivial. If G is not complete, there is a pair of vertices a and b not joined by an edge, and so all remaining vertices form an articulation set. Let U be a minimal articulation set. Let $K_1, K_2, \cdots, K_p$ be the connected components in the subgraph generated by $V(G)-U$, and let G_i be the graph generated by vertices of U and K_i. Using the fact that U is a clique, one shows that $\chi(G)=\max \chi(G_i)$ and $\omega(G)=\max \omega(G_i)$. Since by inductive assumption $\chi(G_i)=\omega(G_i)$ for all i, $\chi(G)$ must equal $\omega(G)$. Q.E.D.

COROLLARY. *Every interval graph is γ-perfect.*

6.5. Multicolorings. An *n-tuple coloring* of a graph G is an assignment of a set $S(x)$ of n different colors to each vertex of G so that if $\{x, y\}$ is an edge of G, then $S(x)$ and $S(y)$ are disjoint. If $\bigcup S(x)$ is a set of k elements, we say that the n-tuple coloring *uses* k colors. Given n, the smallest k so that G has an n-tuple coloring using k colors is called the *n-chromatic number* of G, and is denoted $\chi_n(G)$. This notion has been studied by, among others, Clarke and Jamison (1976), Garey and Johnson (1976), Scott (1975), Stahl (1976), and Chvátal, Garey and Johnson (1976). To give an example, if I_p is the graph consisting of p isolated vertices, then $\chi_n(I_p)=n$. For each vertex can receive the same set of n colors. If G is bipartite and has at least one edge, then $\chi_n(G)=2n$, for each vertex in the same class can receive the same n colors, but the two vertices joined by an edge must receive disjoint sets of colors. To give one final example, Fig. 6.3 shows a 2-tuple coloring of Z_5 using 5 colors.

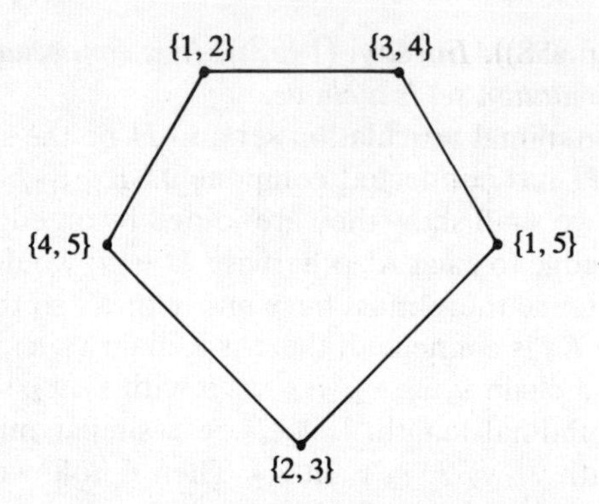

FIG. 6.3. *A 2-tuple coloring of Z_5 using 5 colors.*

The idea of n-tuple coloring arises in the problem of assignment of mobile radio telephone frequencies, which we discussed in § 3.7. We start with a conflict graph whose vertices are zones and whose edges represent conflict between the zones. We wish to assign a band of frequencies $B(i)$ to each zone i so that if there is an edge between i and j, then $B(i) \cap B(j) = \emptyset$. In § 3.7, we thought of these bands as intervals or unions of intervals, and we restricted them to having a certain minimal length. If we think of them all as having the same length or sum of lengths, we may treat them as discrete sets, say of n integers. Then, an assignment of bands which does not cause conflicts corresponds to an n-tuple coloring of the conflict graph. For a further discussion, see Roberts (to appear).

It is interesting to relate the n-chromatic number to the chromatic number. To do so, we follow Harary (1959b) and define a notion of *lexicographic product* $G[H]$ of two graphs. The vertex set of this new graph is the Cartesian product $V(G) \times V(H)$, and there is an edge from (a, b) to (c, d) if and only if either (i) $\{a, c\}$ is an edge of G or (ii) $a = c$ and $\{b, d\}$ is an edge of H. Figure 6.4 gives an example of a lexicographic product of two graphs. As Stahl (1976) observes,

$$\chi_n(G) = \chi(G[K_n]). \tag{6.1}$$

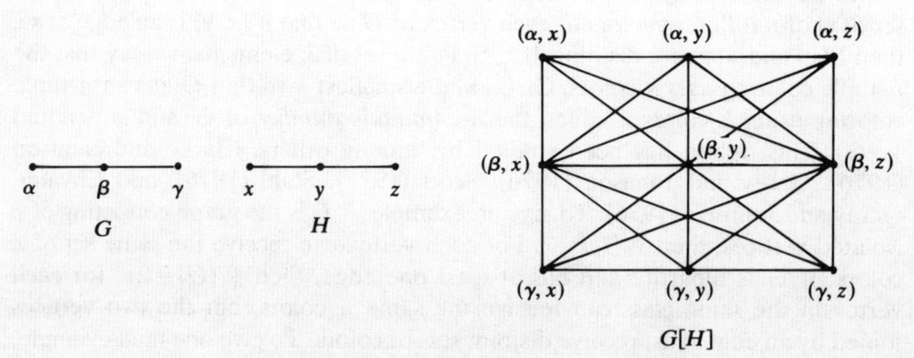

FIG. 6.4. *The lexicographic product of two graphs.*

It will be helpful to make several observations about $G[H]$, some of which will be useful here and some of which will be useful in the next chapter. We shall use the notation $\alpha(G)$ for the size of the largest independent set of vertices in G.

THEOREM 6.4.
1) $G[H]^c = G^c[H^c]$.
2) $\omega(G[H]) = \omega(G)\omega(H)$.
3) $\alpha(G[H]) = \alpha(G)\alpha(H)$.
4) $\chi(G[H]) \leqq \chi(G)\chi(H)$.

Proof. Part 1) is easily verified from the definition and part 3) follows from parts 1) and 2) given the observation that for any graph Γ, $\alpha(\Gamma) = \omega(\Gamma^c)$. To prove part 2), we observe that if K is a maximal clique in G and L is a maximal clique in H, then $K \times L$ is a clique in $G[H]$, and so $\omega(G[H]) \geqq \omega(G)\omega(H)$. Conversely, suppose C is a clique of $G[H]$. Let

$$K = \{a : (a, b) \in C, \text{ some } b\}.$$

Then if $a \neq c$ belong to K, there are b and d so that (a, b) and (c, d) belong to C. Since C is a clique, it follows that $\{a, c\}$ is an edge of G. Hence, K is a clique of G. For each a in K, the number of times a pair of the form (a, x) occurs in C is at most $\omega(H)$. Hence,

$$|C| \leqq |K| \times \omega(H) \leqq \omega(G)\omega(H).$$

Finally, we prove part 4). We shall use the fact that $\chi = \gamma$ and show the result for γ. Suppose $I_1, I_2, \cdots, I_p$ and $J_1, J_2, \cdots, J_q$ are independent sets which partition $V(G)$ and $V(H)$ respectively, and so that $p = \gamma(G)$, $q = \gamma(H)$. Then the sets $I_\alpha \times J_\beta$ are clearly independent in $G[H]$ and partition $V(G[H])$. Q.E.D.

Note that strict inequality may hold in part 4). It is easy to show, for example, that if G is Z_5 and H is K_2, then $\chi(G[H]) = 5$, which is smaller than $\chi(G)\chi(H) = 6$.

THEOREM 6.5. *If G is weakly γ-perfect, then $\chi_n(G) = n\chi(G)$.*

Proof. By Theorem 6.4, part 2),

$$\omega(G[K_n]) = n\omega(G). \tag{6.2}$$

By Theorem 6.4, part 4)

$$\chi(G[K_n]) \leqq n\chi(G). \tag{6.3}$$

Since G is weakly γ-perfect, $\chi(G) = \omega(G)$. Thus, (6.2) and (6.3) imply that

$$\chi(G[K_n]) \leqq n\chi(G) = n\omega(G) = \omega(G[K_n]).$$

Since for any graph, $\chi \geqq \omega$, the theorem follows from (6.1). Q.E.D.

COROLLARY. *If G is weakly γ-perfect, then so is $G[K_n]$.*

The first interesting graph which is not γ-perfect, and hence for which the n-chromatic number is not covered by Theorem 6.5, is Z_5. Stahl (1976) shows that $\chi_n(Z_{2p+1})$ is $2n+1+[(n-1)/p]$, where $[x]$ is the greatest integer less than or equal to x.

6.6. Multichromatic number. Hilton, Rado, and Scott (1973) define the *multichromatic number* $\chi^*(G)$ as follows:

$$\chi^*(G)=\inf\{k/r\colon G \text{ has an } r\text{-tuple coloring using } k \text{ colors}\}. \tag{6.4}$$

Of course, we have

$$\chi^*(G)=\inf\{\chi_r(G)/r\}. \tag{6.5}$$

Stahl (1976) defines an n-tuple coloring of G with k colors to be *efficient* if

$$k/n \leqq \chi_r(G)/r, \quad \text{all } r \geqq 1.$$

A coloring is efficient if the ratio of the number of colors used to the number of colors used per vertex is as small as possible. It is clear that for an efficient coloring, $k/n=\chi^*(G)$. Clarke and Jamison (1976) show that there always is an efficient coloring, i.e., the infimum in (6.5) is always reached. The next theorem follows from Theorem 6.5.

THEOREM 6.6. *If G is weakly γ-perfect, then $\chi^*(G)=\chi(G)$ and G has an efficient* 1*-tuple coloring.*

We shall present an analogous result, which has applications to coding problems, in Chapter 7.

CHAPTER 7

Independence and Domination

In this chapter, we shall discuss applications of the independence number defined briefly in § 6.5 and of another number called the domination number.

7.1. The normal product. Suppose G and H are graphs. Analogous to the lexicographic product defined in § 6.5, we define the *normal product* $G \cdot H$ as follows. The vertices are the pairs in the Cartesian product $V(G) \times V(H)$. There is an edge between (a, b) and (c, d) if and only if one of the following holds:

(i) $\{a, c\} \in E(G)$ and $\{b, d\} \in E(H)$,

(ii) $a = c$ and $\{b, d\} \in E(H)$,

(iii) $b = d$ and $\{a, c\} \in E(G)$.

(The term normal product is used by Berge (1973); another term in use for this is *strong product.*) Figure 7.1 shows a normal product. Notice that it differs from the lexicographic product because the lexicographic product allows edges between (a, b) and (c, d) whenever $\{a, c\} \in E(G)$, even if $b \neq d$ and $\{b, d\} \notin E(H)$.

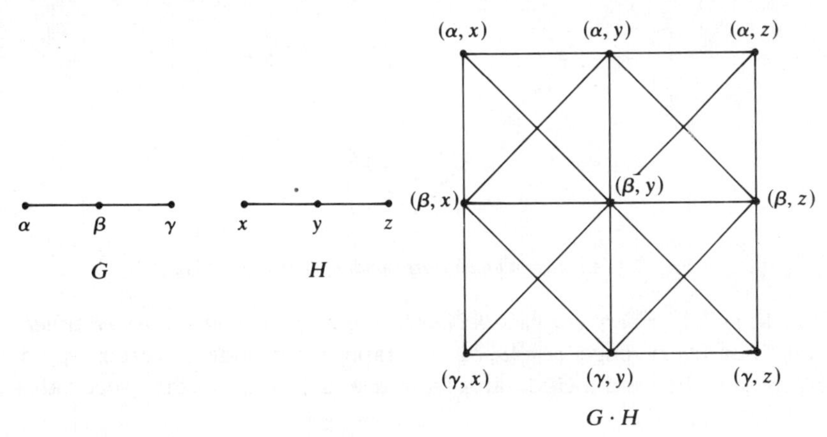

FIG. 7.1. *The normal product of two graphs.*

7.2. The capacity of a noisy channel. In communication theory, a *noisy channel* consists of a *transmission alphabet* T, a *receiving alphabet* R, and information about what letters of T can be *received as* what letters of R. To give an example, suppose a transmitter can emit five signals, a, b, c, d, and e. These letters form the set T. A receiver receives signals α, β, γ, δ, and ε. Unfortunately, because of the noise, confusion is possible: a can be received as either α or β, b as either β or γ, c as either γ or δ, d as either δ or ε, and e as either ε or α. We may summarize this situation by a "receivable as" digraph D, as shown in Fig. 7.2. Corresponding to D is a *confusion graph* G, whose vertices are elements of T and which has an edge between two letters of T if and only if they can be received as the same letter. The confusion graph corresponding to the digraph of Fig. 7.2 is shown in Fig. 7.3. (The reader might wish to compare the confusion graphs to the niche overlap graphs of Chapter 5.)

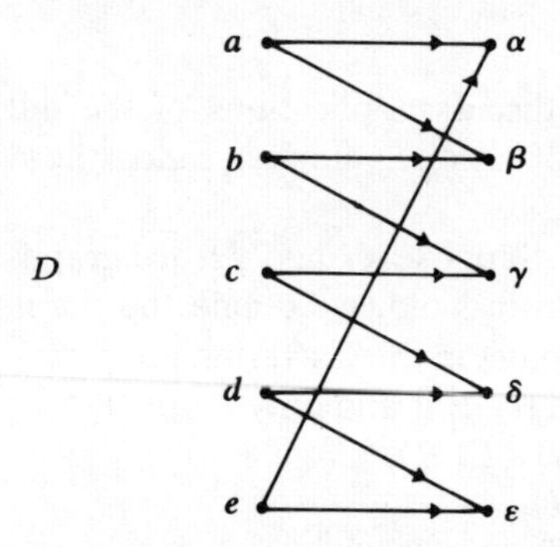

FIG. 7.2. *A "receivable as" digraph.*

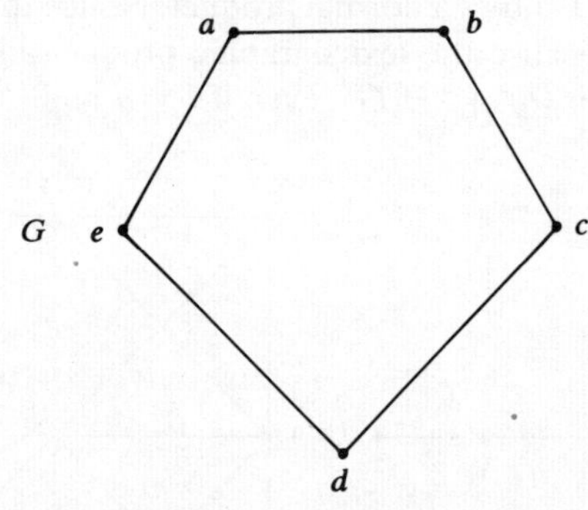

FIG. 7.3. *Confusion graph corresponding to digraph of Fig. 7.2.*

THEOREM 7.1. *Every graph is the confusion graph of some noisy channel.*

Proof. For every edge $\alpha = \{a, b\}$ of graph G, include a vertex x_α in the receiving alphabet and include arcs from a and b to x_α in the "receivable as" digraph D. Q.E.D.

Given a noisy channel, we would like to make errors impossible by choosing a set of signals which can be unambiguously received, i.e., so that no signal in the set is confusable with another signal in the set. This corresponds to choosing an independent set in the confusion graph G. In the graph G of Fig. 7.3, the largest independent set consists of two vertices. Thus, we may choose two such letters, say a and c, and use these as an *unambiguous code alphabet* for sending messages.

Given a fixed noisy channel, we might ask if it is possible to find a larger unambiguous code alphabet. We can find such an alphabet by allowing combinations of letters from the transmission alphabet to form the code alphabet. For example, suppose we consider all possible ordered pairs of elements from T, or strings of two elements from T. Then we can find four such ordered pairs, aa, ac, ca, and cc, none of which can be confused with any of the others. In general, two strings of letters from the transmission alphabet can be confused if and only if they can be received as the same string. In this sense, strings aa and ac cannot be confused, since the only possible strings aa can be received as are $\alpha\alpha$, $\alpha\beta$, $\beta\alpha$, and $\beta\beta$, while the only possible strings ac can be received as are $\alpha\gamma$, $\alpha\delta$, $\beta\gamma$, and $\beta\delta$. We can draw a new confusion graph whose vertices are strings of length two from T. This graph has the following property: strings xy and uv can be confused if and only if one of the following holds:

(i) x and u can be confused and y and v can be confused,
(ii) $x = u$ and y and v can be confused,
(iii) $y = v$ and x and u can be confused.

In terms of the original confusion graph G, the new confusion graph is the normal product $G \cdot G$. (In general, if we have two noisy channels, and consider strings of length two with the first element coming from the first transmission alphabet and the second element from the second transmission alphabet, and if we send the first element of a string over the first channel and the second element over the second, then the resulting confusion graph is the normal product of the two confusion graphs of the two channels.)

If G is the confusion graph of Fig. 7.3, we have already observed that one independent set or unambiguous code alphabet in $G \cdot G$ may be found by using the strings aa, ac, ca, and cc. However, there is a larger independent set, that consisting of the strings aa, bc, ce, db, and ed. In general, if we consider strings of length k from the transmission alphabet of a fixed noisy channel, the confusion graph is the normal product $G^k = G \cdot G \cdot \cdots \cdot G$, with k terms. We search for an independent set in G^k. As before, we let $\alpha(G)$ be the size of the largest independent set in G. We shall study $\alpha(G^k)$.

LEMMA 1. $\alpha(G \cdot H) \geqq \alpha(G)\alpha(H)$.

Proof. If I and J are independent sets in G and H respectively, then the Cartesian product $I \times J$ is independent in $G \cdot H$. Q.E.D.

According to this lemma, $\alpha(G^k) \geqq \alpha(G)^k$. The cost of obtaining a larger independent set in G^k, and hence a larger unambiguous code alphabet, is inefficiency, for the strings in this alphabet are longer. This observation led Shannon (1956) to compensate by considering the number $\sqrt[k]{\alpha(G^k)}$ as a measure of the capacity of the channel to build an unambiguous code alphabet of strings of length k, and to consider the number

$$\alpha^*(G) = \sup_k \sqrt[k]{\alpha(G^k)}.$$

The number $\alpha^*(G)$ is called the *capacity of the graph* or the *zero-error capacity of the channel*, and is usually denoted $c(G)$ in the literature. However, we shall denote it $\alpha^*(G)$ for reasons of comparison with earlier results. If $\alpha^*(G) = \alpha(G)$, then the code cannot be improved by using longer words.

Computation of the capacity of a graph, and determination of graphs whose capacity equals their independence number are both difficult problems, neither of which has been solved. Indeed, even the capacity of the graph $G = Z_5$ which we discussed above was not known precisely until Lovász (1977) showed that it equals $\sqrt{5}$, i.e., that

$$c(Z_5) = \sqrt{\alpha(Z_5^2)}.$$

Meanwhile, as of this writing, $c(Z_7)$ remains unknown. A general upper bound for capacity of an arbitrary graph is given by Rosenfeld (1967).

We shall give some sufficient conditions for $\alpha^*(G)$ to equal $\alpha(G)$. Let $\theta(G)$ be the minimum number of (not necessarily dominant) cliques which partition $V(G)$. For example, in the graph of Fig. 7.4, $\theta(G) = 2$ and two cliques which suffice are $\{a, b, c\}$ and $\{d, e\}$. Note that in this example, $\alpha(G) = \theta(G)$.

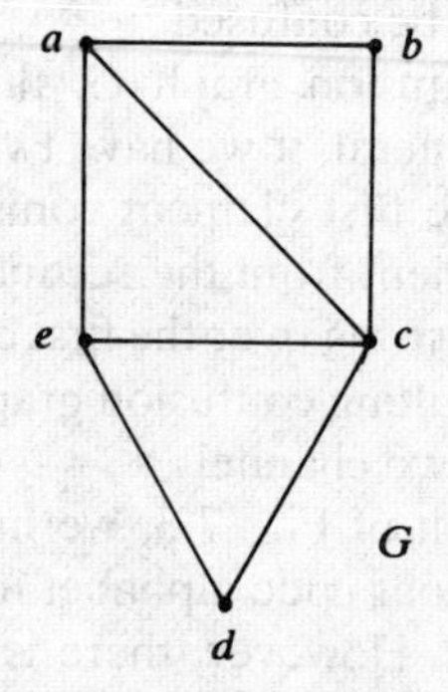

FIG. 7.4. *A graph with* $\theta(G) = 2$.

LEMMA 2. $\alpha(G) \leqq \theta(G)$.

Proof. If I is an independent set of G, each vertex must be in a different clique in any partition. Q.E.D.

Of course, it is easy to give examples where $\alpha < \theta$. The graph Z_5 has $\alpha = 2$ while $\theta = 3$ (two edges and a single vertex are required). We shall show that the condition $\alpha = \theta$ is sufficient for the conclusion that $\alpha^* = \alpha$.

LEMMA 3. $\theta(G \cdot H) \leqq \theta(G)\theta(H)$.

Proof. Suppose $K_1, K_2, \cdots, K_r$ are cliques partitioning $V(G)$ and $L_1, L_2, \cdots, L_s$ are cliques partitioning $V(H)$. Then $\{K_i \times L_j\}$ is a set of cliques of $G \cdot H$ partitioning $V(G \cdot H)$. Q.E.D.

We shall call a graph G *weakly α-perfect* if $\alpha(G) = \theta(G)$ and *α-perfect* if every generated subgraph of G is weakly α-perfect.

THEOREM 7.2 (Shannon (1956)). *If G is weakly α-perfect, then $\alpha^*(G) = \alpha(G)$.*

Proof. By Lemmas 1, 2, and 3,

$$\alpha(G)^k \leqq \alpha(G^k) \leqq \theta(G^k) \leqq \theta(G)^k.$$

Since G is weakly α-perfect, the first and last terms are equal. Q.E.D.

Let us now observe that there are many examples of weakly α-perfect graphs. For, by Lovász' theorem (Theorem 6.2), a graph G is γ-perfect if and only if its complement is γ-perfect. Now it is clear that

$$\alpha(G) = \omega(G^c).$$

Moreover,

$$\theta(G)=\chi(G^c),$$

for the chromatic number is the smallest number of independent sets which partition the vertices of a graph. It follows from Lovász' theorem that for every graph G, G is α-perfect if and only if G is γ-perfect. In particular, we conclude that all rigid circuit graphs, and hence all interval graphs, being γ-perfect, have $\alpha^*=\alpha$. Thus, channels whose confusion graphs are rigid circuit graphs, in particular interval graphs, cannot be improved by using longer strings. Channels with interval graph confusion graphs arise when a signal is determined by its modulation frequency and two signals are confusable if and only if their corresponding intervals overlap. This is the situation called *linear noise.* Hence, if noise is linear, then, as Berge (1973) points out, a code cannot be improved by using longer strings.

We should also note that Theorem 7.2 is analogous to the result (Theorem 6.6) which says that if G is weakly γ-perfect, then $\chi^*(G)=\chi(G)$.

We conclude this section by showing that for every positive number k, there is a graph G_k so that $\alpha^*(G_k)\geqq k\alpha(G_k)$. Thus, capacity can be arbitrarily larger than the largest independent set. We follow Rosenfeld (1970). A graph G is said to be *self-complemented* if G is isomorphic to G^c. Z_5 is an example of such a graph. The isomorphism is shown in Fig. 7.5.

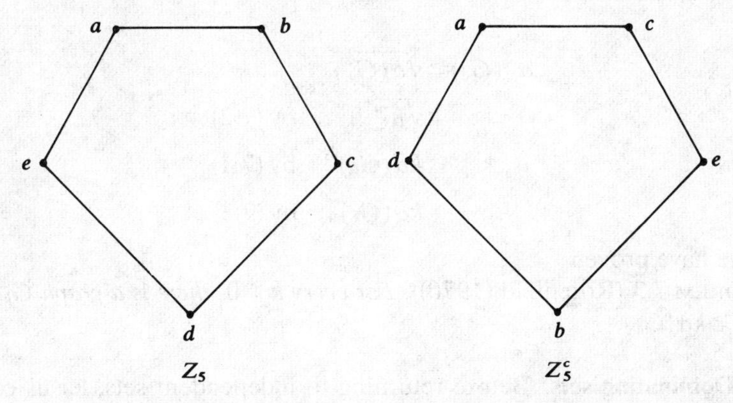

FIG. 7.5. *Z_5 is self-complemented, with the isomorphism from Z_5 to Z_5^c given by $a\to a$, $b\to c$, $c\to e$, $d\to b$, $e\to d$.*

LEMMA 4. *If G is self-complemented, then $\alpha(G^2)\geqq|V(G)|$.*

Proof. Let

$$I=\{(x,x):x\in V(G)\}.$$

Then I is independent in $G\cdot G^c$. Since $G\cdot G$ is isomorphic to $G\cdot G^c$, there is an independent set in G^2 of at least $|I|$ elements. Q.E.D.

It is interesting to note that for Z_5, the independent set I of the proof of Lemma 4 corresponds to the independent set *aa, bc, ce, db,* and *ed* in $G\cdot G$

under the isomorphism of Fig. 7.5. This independent set is exactly the one we identified earlier.

The next lemma concerns the lexicographic product $G[H]$ defined in § 6.5.

LEMMA 5. *If G and H are self-complemented, then $G[H]$ is self-complemented.*

Proof. Use Theorem 6.4, part 1). Q.E.D.

Now let G_0 be a self-complemented graph with $\alpha(G_0)^2 < |V(G_0)| = n$, for example, $G_0 = Z_5$. Given k, there is an $m > 0$ so that

$$n^m \geqq k^2 \alpha(G_0)^{2m},$$

and hence

$$\sqrt{n^m} \geqq k\alpha(G_0)^m. \tag{7.1}$$

Let $G_k = G_0[G_0[G_0 \cdots]]$, where the lexicographic product is taken m times. By Lemma 5, G_k is self-complemented. Hence, by Lemma 4,

$$\alpha(G_k^2) \geqq |V(G_k)| = n^m. \tag{7.2}$$

By Theorem 6.4, part 3),

$$\alpha(G_k) = \alpha(G_0)^m. \tag{7.3}$$

Now

$$\begin{aligned} \alpha^*(G_k) &\geqq \sqrt{\alpha(G_k^2)} \\ &\geqq \sqrt{n^m} && \text{by (7.2)} \\ &\geqq k\alpha(G_0)^m && \text{by (7.1)} \\ &= k\alpha(G_k) && \text{by (7.3).} \end{aligned}$$

Thus, we have proven

THEOREM 7.3 (Rosenfeld (1970)). *For every $k > 0$, there is a graph G_k so that $\alpha^*(G_k) \geqq k\alpha(G_k)$.*

7.3. Dominating sets. Before returning to independent sets, let us consider another type of set. Let D be a digraph. A set B of vertices is called a *dominating set* if whenever x is not in B, there is some y in B so that (y, x) is an arc of D. To give an example, in the digraph of Fig. 7.6, the set $\{a, c\}$ is a dominating set. The *domination number* of a digraph is the size of the smallest dominating set.

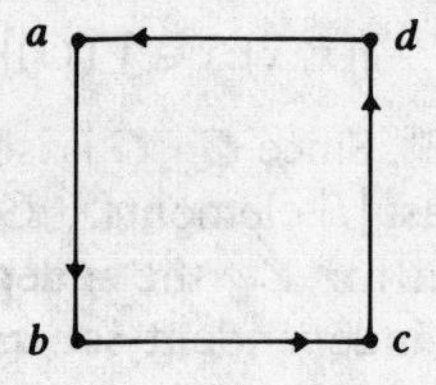

FIG. 7.6. *The set $\{a, c\}$ is a dominating set and a stable set.*

Dominating sets arise in a number of related problems. Berge (1973) talks about the problem of locating radar stations. A number of strategic locations are to be kept under observation. However, it is desired to put radar for the observation process at as few of these locations as possible. How can we determine a set of locations in which to place the radar stations? We draw a digraph D with vertex set the locations in question. We draw an arc from x to y if it is possible to observe y from a radar station at x. An acceptable set of locations in which to place radar stations corresponds to a dominating set in this digraph, and we wish to find such a set of minimum size.

A similar problem arises in nuclear power plants. Here, we have various locations, and we draw an arc from location x to location y if it is possible for a watchman stationed at x to observe a warning light located at y. How many guards are needed to observe all of the warning lights, and where should they be located? The answer again corresponds to a minimum dominating set.

Similarly, suppose we have communication links in existence between cities, and we want to set up transmitting stations at some of the cities so that every city can receive a message from at least one of the transmitting stations. Again, as Liu (1968) points out, we are searching for a dominating set.

Finally, dominating sets have an interesting application in voting situations, as pointed out by Harary, Norman and Cartwright (1965). Suppose a group is trying to form a responsive committee to represent it. Let each member of the group designate that individual who he feels best understands his needs and would best represent his views. Let the group members be vertices of a digraph and draw an arc from vertex x to vertex y if x was designated by y. Then a minimum dominating set would make a representative committee.

The notion of dominating set can be modified in various ways. For example, we might be willing to find a *k-dominating set*, or a *k-cover*, a set C of vertices in a digraph D so that every vertex of D can be reached by a path of length at most k from a vertex of C. Or, we might have a designated subset S of $V(D)$, and search for a set of vertices which dominates every vertex of S. The latter problem is clearly of importance in modifications of the radar station and nuclear power plant observation problems. The former problem might be important if the vertices are locations in a city and the k-cover we seek is a set of locations in which to put emergency services such as police stations or fire stations. A famous result about tournaments (due to Landau (1955)) says that if x is a winner of a round–robin tournament, a player having a maximum number of victories, then $\{x\}$ is a 2-dominating set in the digraph which has an arc from player a to player b if a beats b.

7.4. Stable sets. The notion of independent set makes sense for a digraph as well as a graph: a set of vertices in a digraph D is called *independent* if there is no arc joining any two vertices in the set. A set of vertices which is at the same time an independent set and a dominating set is called a *stable set* or a *kernel* of the digraph. The set $\{a, c\}$ in the digraph of Fig. 7.6 is an example of a stable set.

Stable sets were introduced into the theory of games by von Neumann and Morgenstern (1944), where they were used to define possible solutions to a game. In particular, suppose the vertices of a digraph represent possible outcomes of a game, and we draw an arc from vertex x to vertex y if and only if some group of players *effectively prefers* x to y, i.e., not only prefers x to y, but has sufficient power to make its preference effective (this term is defined more precisely in game theory). Then von Neumann and Morgenstern seek a set of outcomes which has the properties that no outcome in the set is effectively preferred to any other outcome in the set (independence) and that for every outcome y not in the set there is an outcome x in the set so that x is effectively preferred to y (domination). Modern-day game theorists apply stable sets as "solutions" to games representing bargaining situations such as in voting, in the economic marketplace, and in international relations (oil cartels, deterrence, disarmament, etc.).

Liu (1968) points out that stable sets are also of interest in cross-referencing in computerized automatic library systems. Draw a digraph which has as vertex set the books in the system, and include an arc from book x to book y if x refers to y. A stable set in this digraph corresponds to a set of books from which it is possible to branch out to all other books in the library, and which has the property that no book in the set refers to any other book in the set.

Unfortunately, not every digraph has a stable set. For example, a cycle of length 3 does not have such a set. Other digraphs, such as that of Fig. 7.6, have more than one stable set. Thus, there has been considerable interest in studying the existence and uniqueness of stable sets. We shall state some results here, following Berge (1973) for those theorems for which credit is not given. See also Roberts (1976a) for a discussion of stable sets and their applications to game theory.

THEOREM 7.4. *Every stable set is a maximal independent set and a minimal dominating set.*

THEOREM 7.5. *Every symmetric digraph has a stable set, and in such a digraph a set is stable if and only if it is a maximal independent set.*

THEOREM 7.6 (von Neumann and Morgenstern (1944)). *Every acyclic digraph has a unique stable set.*

THEOREM 7.7 (Harary, Norman and Cartwright (1965)). *Every strongly connected digraph D consisting of more than one vertex and having no odd cycles has at least two stable sets.*

THEOREM 7.8 (Richardson (1946)). *Every digraph with no odd cycles has a stable set.*

To illustrate some of these results, let us note that the set $\{a, b, c, d, e\}$ is the unique stable set in the digraph of Fig. 7.2. It is clearly a maximal independent set and a minimal dominating set. This illustrates Theorems 7.4 and 7.6. If the graph G of Fig. 7.4 is considered a symmetric digraph, then the sets $\{a, d\}$, $\{b, d\}$, $\{b, e\}$, and $\{c\}$ are the maximal independent sets. These are also all the stable sets, thus illustrating Theorem 7.5. The digraph of Fig. 7.6 illustrates Theorem 7.7, for there are two stable sets, the sets $\{a, c\}$ and $\{b, d\}$.

CHAPTER 8

Applications of Eulerian Chains and Paths

8.1. Existence theorems. A chain (path) in a graph G (digraph D) is called *eulerian* if it uses every edge of G (arc of D) once and only once. Eulerian chains arose from the Königsberg bridge problem, which asked whether the townspeople in Königsberg could traverse a series of bridges, going over each once and only once, and returning to the starting point. In the course of showing this was impossible, Euler produced techniques which gave birth to graph theory. In today's world, the notions of eulerian chain and path are applied to such problems as routing street-sweeping and snow-removal vehicles, untangling genetic information, and designing telecommunications systems. We shall investigate these applications in this chapter.

It will be useful in this chapter to allow a graph or digraph to have more than one edge or arc between vertices. We shall use the terms *multigraph* and *multidigraph* to make it clear that multiple edges and arcs are allowed. Loops will also be allowed. The notions of chain, path, etc. are unchanged, and we still define a chain or path as *eulerian* if it goes through each edge or arc once and only once.

If a multigraph has an eulerian closed chain, then it must be connected up to isolated vertices. Moreover, since an eulerian closed chain must leave every vertex as often as it enters, each vertex must have even degree, where the *degree* of a vertex is the number of edges joining it.

THEOREM 8.1. *A multigraph G has an eulerian closed chain if and only if G is connected up to isolated vertices and every vertex has even degree.*

Figure 8.1 illustrates this theorem. Every vertex of the multigraph G has even degree. An eulerian closed chain is the following: $a, b, c, b, c, d, a, e, d, f, a$. For a proof of Theorem 8.1, see for example Harary (1969).

THEOREM 8.2. *A multigraph G has an eulerian chain if and only if G is connected up to isolated vertices and the number of vertices of odd degree is either* 0 *or* 2.

To illustrate this theorem, note that the multigraph of Fig. 8.2 has an eulerian chain a, b, c, d, a, c, d, but no eulerian closed chain since vertices a and d have odd degree.

If D is a multidigraph the *outdegree* and *indegree* of a vertex are defined to be the number of outgoing and incoming arcs respectively. We abbreviate these notions by od and id respectively. If a multidigraph has an eulerian closed path, it must up to isolated vertices be *weakly connected*, i.e., if direction is disregarded, each vertex is reachable from each other vertex.

THEOREM 8.3 (Good (1947)). *A multidigraph D has an eulerian closed path if* *and only if D is weakly connected up to isolated vertices and for every vertex, the indegree equals the outdegree.*

THEOREM 8.4 (Good (1947)). *A multidigraph D has an eulerian path if and only if D is weakly connected up to isolated vertices and for all vertices with the possible exception of two, the indegree equals the outdegree, and for at most two vertices, the indegree and outdegree differ by one.*

The first multidigraph of Fig. 8.3 does not have an eulerian path since there are vertices where the indegree and outdegree differ by two. The second multidigraph does not have an eulerian path because there are four vertices where the indegree and outdegree differ by one. Notice that since the sum of the indegrees equals the sum of the outdegrees, it follows from Theorem 8.4 that if there is one exceptional vertex, there will be two, and one will have $\mathrm{id}=\mathrm{od}+1$, while the other has $\mathrm{id}=\mathrm{od}-1$. Note that these theorems hold if there are loops. A loop will contribute indegree and outdegree of 1 to the given vertex, and so loops make no difference as to existence of eulerian chains, paths, closed chains, and closed paths.

8.2. The transportation problem. The following problem arises in a variety of applications, and is called the *transportation problem.* We shall refer to it in applications of eulerian paths. Let there be a certain number of warehouses and a certain number of markets. Let a_{ij} be the cost of transporting one unit of a commodity from warehouse i to market j. Let x_i be the number of units of the commodity at warehouse i and y_j be the number of units of the commodity required at market j. Determine how much of the stock at each warehouse should be shipped to each market so as to minimize the total transportation cost. This problem has been extensively studied, and there are very practical algorithms for its solution. We shall not present any such algorithm here, but simply refer the reader to such texts as Hillier and Lieberman (1974) or Wagner (1975). In the next section, we shall show how to make use of algorithms for solving the transportation problem in finding certain kinds of eulerian closed paths.

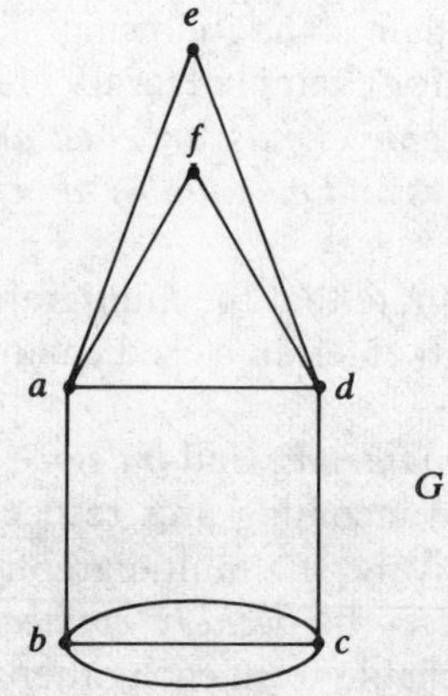

FIG. 8.1. *A multigraph with an eulerian closed chain.*

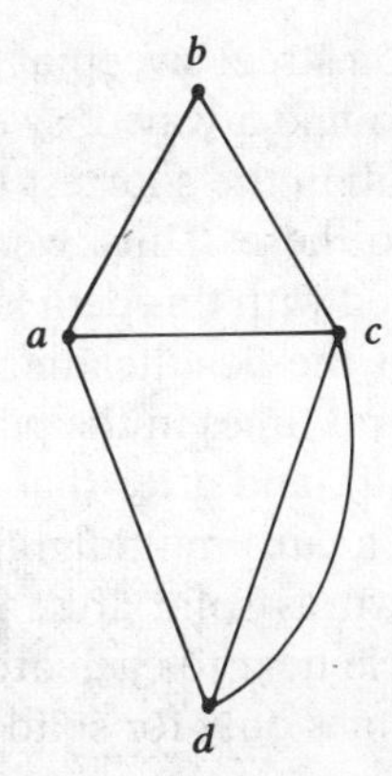

FIG. 8.2. *A multigraph with an eulerian chain but no eulerian closed chain.*

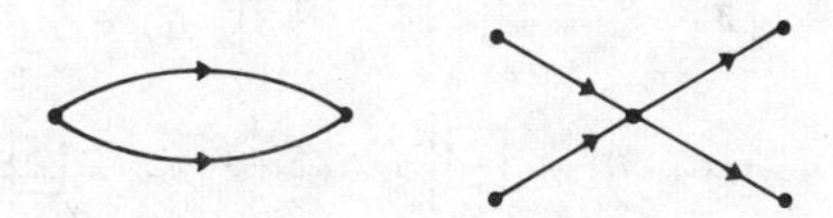

FIG. 8.3. *Two multidigraphs without eulerian paths.*

8.3. Street-sweeping. Millions of dollars are spent each year by municipalities in performing services such as street-sweeping, snow-removal, etc. We shall show how the notion of eulerian closed path enters into the determination of optimal routes for street-sweepers and snow-removers. Our discussion follows Tucker and Bodin (1976). For a similar discussion, with examples from the city of Zurich, see Liebling (1970). We shall speak in the language of street-sweeping for concreteness.

In general, we shall draw a multidigraph, the *curb multidigraph*, to correspond to the streets of a city. The vertices are street corners, and the arcs correspond to curbs. Thus, there is an arc from street corner x to street corner y if there is a curb which can be traveled along from x to y. We obtain a multidigraph because in a one-way street, there are two curbs which can only be swept by going in the same direction along the street. Each arc of this multidigraph has two numbers associated with it, one giving the amount of time required to sweep the corresponding curb and the other giving the amount of time required to follow along the arc without sweeping, the so-called *deadheading time.*

Now in any given period of the day in large cities such as New York, certain curbs are kept free of parked cars in order to allow for street sweeping. The curbs which are free during a given period of time define a subgraph (subdigraph) of the curb multidigraph. This subgraph is called the *sweep subgraph.* (It is an interesting question in its own right to determine how to choose these subgraphs so as to save money on street sweeping. However, we shall not discuss that problem.) We would like to find a way of sweeping every curb in the sweep subgraph and completing the job in the shortest possible time. We wish to start out from a garage and return to there. Thus, we seek a closed path in the curb multidigraph. The time associated with this path is the sum of the sweeping times over arcs swept plus the sum of the deadheading times over arcs traversed but not swept. If an arc is used several times in the path, it is considered to be swept the first time (if it is swept at all), and after that the deadheading time is used. Figure 8.4 gives an example of a curb multidigraph. The arcs in the designated subgraph to be swept are shown as solid arcs, the other arcs as dashed. The sweeping time is shown in a circle next to each arc, and the deadheading time in a square. (We show sweeping times only for solid arcs.)[16]

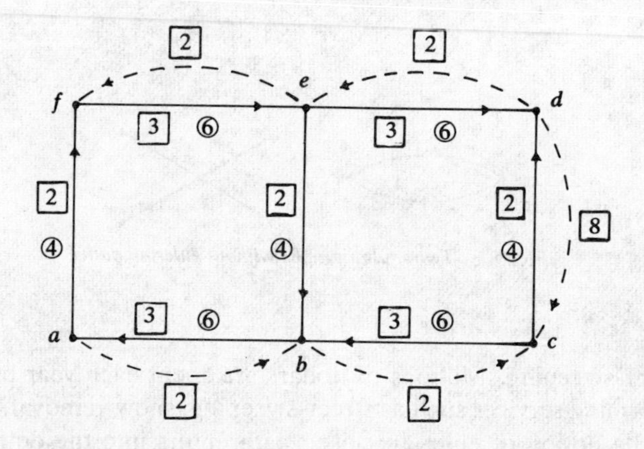

vertex i	od(i)	id(i)	$d(i)$
	in sweep subgraph		
a	1	1	0
b	1	2	−1
c	2	0	2
d	0	2	−2
e	2	1	1
f	1	1	0

		2	1	y_j
x_i	warehouses ↓	c	e	← markets
1	b	2	6	
2	d	6	2	$= (a_{ij})$

FIG. 8.4. *A curb multidigraph, with solid arcs representing the sweep subgraph, sweeping times given in circles, and deadheading times given in squares. Also shown are the degrees of vertices in the sweep subgraph and the transportation problem whose solution gives rise to an optimal eulerian path.*

[16] Our discussion omits time delays associated with turns—some turns may take longer than others—and other delays. These delays can be introduced by defining a system of penalties for turns and other delays as well as for time spent sweeping or deadheading. See Tucker and Bodin (1976) for details.

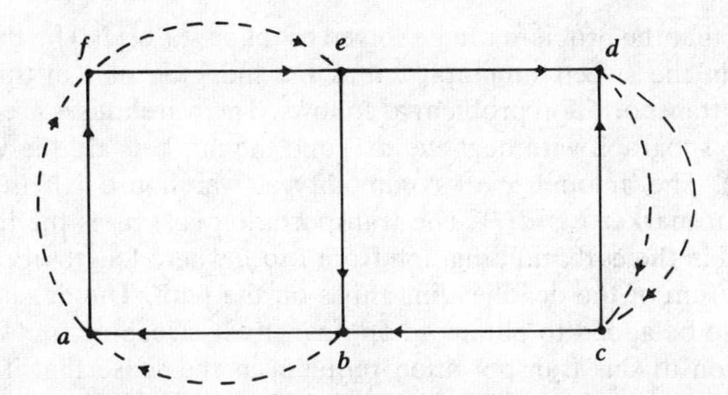

FIG. 8.5. *A multidigraph with an eulerian closed path obtained by adding dashed arcs to the sweep subgraph of Fig.* 8.4.

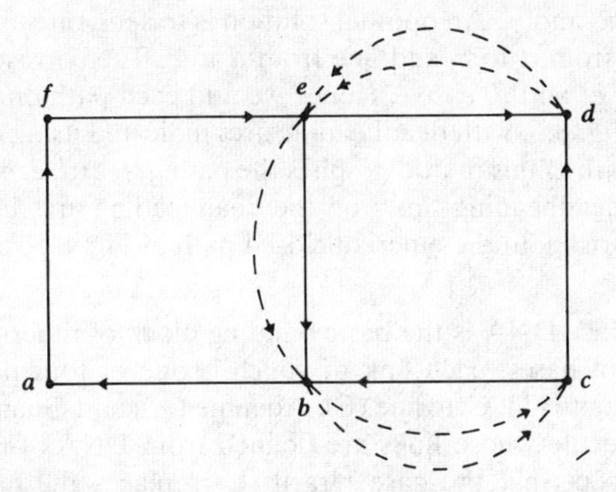

FIG. 8.6. *A multidigraph obtained from the sweep subgraph of Fig.* 8.4 *by adding paths corresponding to the solution of the transportation problem.*

Notice that if there is an eulerian closed path in the sweep subgraph, then that closed path must be an optimal solution. The problem is to handle the situation where there is no such path. In our example, there are four vertices where the indegrees and outdegrees (shown in Fig. 8.4) are unequal in the sweep subgraph. Hence, there is no eulerian closed path.

Any closed path of the curb multidigraph which uses each arc of the sweep subgraph at least once can be looked at as obtained from the sweep subgraph by adding arcs to obtain a multidigraph with an eulerian closed path. The added arcs may be arcs in the curb multidigraph not in the sweep subgraph, or arcs in the sweep subgraph which are used again. A given arc may be added more than once. To give an example, consider in Fig. 8.4 the closed path *f, e, b, a, f, e, d, c, d, c, b, a, f*. This corresponds to an eulerian closed path in the multidigraph shown in Fig. 8.5. The arcs added (deadheading arcs) are dashed. We would like to add arcs to the sweep subgraph so that in the resulting multidigraph, every vertex has outdegree equal to indegree. We would also like to do this so that the sum of the deadheading times on the arcs added is minimized. Tucker and Bodin

(1976) show that the problem can be solved as follows. Let $d(i)$ be the outdegree of vertex i in the sweep subgraph minus the indegree of i in this subgraph. Formulate a transportation problem as follows. The warehouses are the vertices of the sweep subgraph with negative $d(i)$ and the markets are the vertices with positive $d(i)$. The amount x_i of commodity at warehouse i is $|d(i)|$ and the amount y_j for market j is $d(j)$. The transportation cost a_{ij} is the length of the shortest path in the curb multidigraph from i to j, where length of path is taken to mean the sum of the deadheading times on the path. The deadheading arcs which need to be added to obtain an optimal street-sweeping route correspond to the solution to this transportation problem in the sense that if b_{ij} units of commodity are shipped from i to j in the solution, then the shortest path from i to j is included b_{ij} times.

In our example, the degrees $d(i)$ are shown in Fig. 8.4, as are the matrix (a_{ij}) and the amounts x_i and y_j. An optimal solution is to send one unit of commodity from b to c, one from d to e, and one from d to c. The corresponding shortest paths are b, c; d, e; and d, e, b, c. Hence we add each path once, obtaining the multidigraph of Fig. 8.6, with deadheading arcs indicated using dashed lines. An eulerian closed path in this multidigraph is the path $a, f, e, d, e, b, c, b, c, d, e, b, a$. The sum of the deadheading times on the deadheading arcs in this path is 10. By way of comparison, in the eulerian closed path of Fig. 8.5, this sum is 24.

8.4. RNA chains. DNA is the basic building block of inheritance. DNA is a chain consisting of bases, each link of which is one of four possible chemical compounds: Thymine (T), Cytosine (C), Adenine (A), and Guanine (G). RNA is a messenger molecule whose links are defined from DNA. The possible bases are the same except that the base Uracil (U) replaces the base Thymine. A sequence of bases encodes certain genetic information. It is an elementary problem of combinatorics to count the number of possible RNA chains with certain link makeup. For example, the number of such chains with 3 C's, 2 U's and 2 A's is obtained as follows. We have seven positions, and we choose 3 of these for C. This can be done in $\binom{7}{3}$ ways. We choose 2 of the remaining 4 positions for U. This can be done in $\binom{4}{2}$ ways. Finally, we use the remaining positions for A. Altogether, we have

$$\binom{7}{3}\binom{4}{2}=\frac{7!}{3!2!2!}=210$$

such chains. One such chain is CUACUAC. In general, the number of chains with k_1 C's, k_2 U's, k_3 A's, and k_4 G's is given by

$$\frac{(k_1+k_2+k_3+k_4)!}{k_1!k_2!k_3!k_4!}. \tag{8.1}$$

This can be quite a large number of chains.

Sometimes we can learn much more about what an RNA chain looks like by considering the break-up of the chain after certain enzymes are applied. Our discussion will follow Mosimann (1968) and Hutchinson (1969). Some enzymes break up an RNA chain after each G link and others break up the chain after each U link and each C link. For example, suppose we have the chain

GAUGGACC.

Applying (8.1), we see that there are $8!/(2!1!2!3!) = 1680$ chains with the same link makeup. Digest by the G and U, C enzymes leads to the following fragments:

G: G, AUG, G, ACC

U, C: GAU, GGAC, C.

How many RNA chains are there with the given U, C fragments, if we don't know the order in which the fragments occur? The answer is that there are 3! such chains, for the fragments may occur in any order. As for the G fragments, there are not 4! distinguishable chains with the given fragments, since there are two indistinguishable fragments. The number of chains with these fragments is therefore $4!/2! = 12$. In fact, there are fewer chains with these G fragments, since we can pick out which fragment came last. The fragment ACC could be a G fragment only if it ended up the chain. Thus, the chain can only start with the three G fragments G, AUG, and G. There are now only $3!/2! = 3$ possible ways to order these beginning fragments and hence only 3 possible chains with the given G fragments:

GAUGGACC

GGAUGACC

AUGGGACC.

Thus, knowing either the U, C fragments or the G fragments significantly lowers the number of possible chains. However, we know *both* sets of fragments. Of the possible chains we have identified as having the proper G fragments, it is easy to check that only the first has the proper U, C fragments. For the second has GGAU as a U, C fragment and the third has AU, neither of which appear among the given U, C fragments. Thus we have been able to uncover the original RNA chain from all 1680 chains with the given link makeup by applying appropriate enzymes. We shall see how to perform this procedure using eulerian closed paths in a certain multidigraph.

Let us illustrate the procedure by starting with an example of an unknown RNA chain. Suppose after application of the G and U, C enzymes, we find the following fragments:

G: AUCG, G, CCG, AG, UAC

U, C: C, C, C, GAU, GGAGU, AC.

Let us begin by further breaking down each fragment, after each G, U, or C. For example, the fragment AUCG gets broken into AU · C · G. Each piece is called an *extended base*, and all but the first and last extended bases are called *interior extended bases.* We will first see how to discover both the beginning and end of the chain. We make two lists, one giving all interior extended bases of all fragments of both digests, and one giving all fragments consisting of one extended base. We obtain the following:

Interior extended bases: C, C, G, AG

Fragments consisting of one extended base: G, AG, C, C, C, AC.

Note that the last two interior extended bases come from the fragment GGAGU. By comparing our two lists, we observe that there are two bases on the second list which are not on the first, namely C and AC. This will always be the case. Moreover, it is not hard to prove that one of these will be the first extended base of the RNA chain and the other will be the last. How do we tell which is last? The answer is that one of these will always be from an *abnormal fragment*, namely, it will be the last extended base of a G fragment not ending in G or a U, C fragment not ending in U or C. In this case, AC is the last extended base of the abnormal G fragment UAC. Hence, we know that the chain begins in C and ends in AC.

We now build a multidigraph as follows. Whenever there is a normal fragment with more than one extended base, we use the first and last extended bases of the fragment as vertices and we draw an arc from the first to the last, labeled with the name of the corresponding fragment. Figure 8.7 illustrates this procedure. For example, we have included an arc from AU to G labeled with the name of the corresponding fragment AUCG. Notice that there might be several arcs from a given extended base to another if there are several fragments beginning and ending with the given extended bases. For example, if there were fragments AUCG and AUAUG, we would put two arcs from vertex AU to vertex G, one labeled with each fragment. Finally, we add one additional arc to this multidigraph. This arc is obtained by identifying the longest abnormal fragment—here there is just one, namely UAC—and drawing an arc from the first extended base in this abnormal fragment to the first extended base in the chain. Here, we add an arc from U to C. We label this arc differently, by marking it X^*Y, where X is the longest abnormal fragment, * is a symbol marking this as a special fragment, and Y is the first extended base in the chain. Hence, in our example, we label the arc from U to C by UAC*C. Every possible RNA chain with the given C and U, C fragments can now be identified from this multidigraph. Each such chain corresponds to an eulerian closed path which ends with the special fragment X^*Y. In our example, the only such eulerian closed path goes from C to G to AU to G to U to C. By using the corresponding arc labeling, we obtain the chain

CCGAUCGGAGUAC.

The reader can check that this chain has the proper G and U, C fragments.

Notice that we are not claiming that there is a unique eulerian closed path in the multidigraph, or that there is a unique RNA chain with the given G and U, C fragments. Indeed, it is easy to give examples of RNA chains with ambiguous digests, i.e., so that it is impossible to recover the chain uniquely from its G and U, C fragments. For further information on the theory of digests by enzymes, mostly from an algebraic point of view, see for example Mosimann, et al. (1966).

It should be remarked that current work of great societal import is based on the idea of breaking up DNA chains with enzymes and splicing together fragments from different species, e.g., from humans and bacteria. This work on *recombinant* DNA opens up great possibilities such as in the production of insulin. However, it opens up the possibility of grave danger from the creation of potentially harmful individuals of previously unknown genetic makeup. Techniques of the type discussed above allow for the possibility of identifying all potential recombinants after a given enzyme break-up.

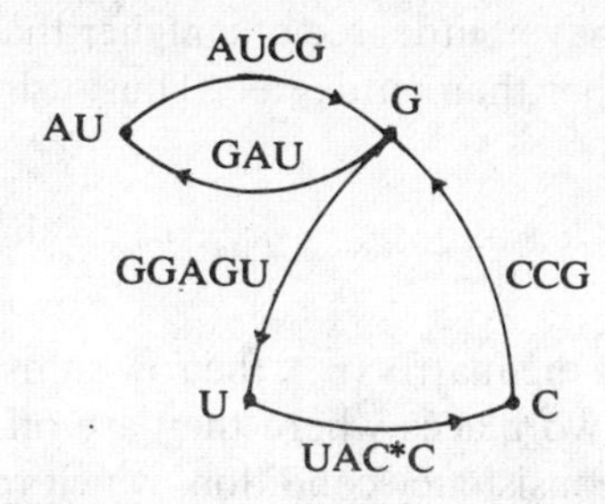

FIG. 8.7. *A multidigraph from a complete digest by* G *and* U, C enzymes.

8.5. More on eulerian closed paths, DNA, and coding. Hutchinson and Wilf (1975) treat a DNA or RNA molecule as a word, with bases (not extended bases) as the letters. They make the simplifying assumption that all of the information is carried only in the number of letters of each type and in the frequency of ordered pairs of letters, i.e., the frequency with which one letter follows a second. They then ask the following question: given nonnegative integers v_i, v_{ij}, $i, j = 1, 2, \cdots, n$, can a word be made from an alphabet of n letters with the ith letter occurring v_i times and with i followed by j v_{ij} times? If so, what are all such words? This question is of general interest in the construction of codes. We follow Hutchinson and Wilf's solution.

To give an example, suppose $v_1 = 2$, $v_2 = v_3 = 1$ and v_{ij} is given by the following matrix

$$(v_{ij}) = \begin{pmatrix} 0 & 1 & 0 \\ 0 & 0 & 1 \\ 1 & 0 & 0 \end{pmatrix}. \tag{8.2}$$

Then one word which has the prescribed pattern is $ABCA$, if A corresponds to the first letter, B to the second, and C to the third. To give a second example, suppose $v_1 = 2$, $v_2 = 4$, $v_3 = 3$, and

$$(8.3) \qquad (v_{ij}) = \begin{pmatrix} 0 & 0 & 2 \\ 2 & 1 & 1 \\ 0 & 2 & 0 \end{pmatrix}.$$

One word which has the prescribed pattern is *BBCBACBAC*.

To analyze our problem, let us draw a multidigraph D with vertices the n letters $A_1, A_2, \cdots, A_n$, and with v_{ij} arcs from A_i to A_j. Loops are allowed. The multidigraphs corresponding to the matrices of (8.2) and (8.3) are shown in Figs. 8.8 and 8.9 respectively. Let us suppose $w = A_{i_1}, A_{i_2}, \cdots, A_{i_q}$ is a solution word. Then it is clear that w corresponds to an eulerian path in the multidigraph D which begins at A_{i_1} and ends at A_{i_q}. It is easy to see this for the two solution words we have given for our two examples. It follows that if there is a solution word, then D must be weakly connected up to isolated vertices. We consider first the case where $i_1 \neq i_q$. For every $i \neq i_1, i_q$, we have indegree at A_i equal to outdegree. For $i = i_1$, we have outdegree one higher than indegree and for $i = i_q$, we have indegree one higher than outdegree. Thus, using δ to be the Kronecker delta, we have

$$(8.4) \qquad \sum_{k=1}^{n} v_{ik} = \sum_{k=1}^{n} v_{ki} + \delta_{ii_1} - \delta_{ii_q}, \quad \text{all } i.$$

This condition says that in the matrix (v_{ij}), the row sums equal the corresponding column sums, except in two places where they are off by one in the indicated manner. We also have a consistency condition, which relates the v_i to the v_{ij}:

$$(8.5) \qquad v_i = \sum_{k=1}^{n} v_{ki} + \delta_{ii_1}, \quad \text{all } i.$$

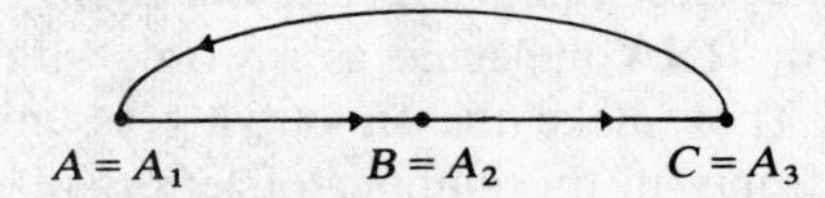

FIG. 8.8. *Multidigraph corresponding to* (8.2).

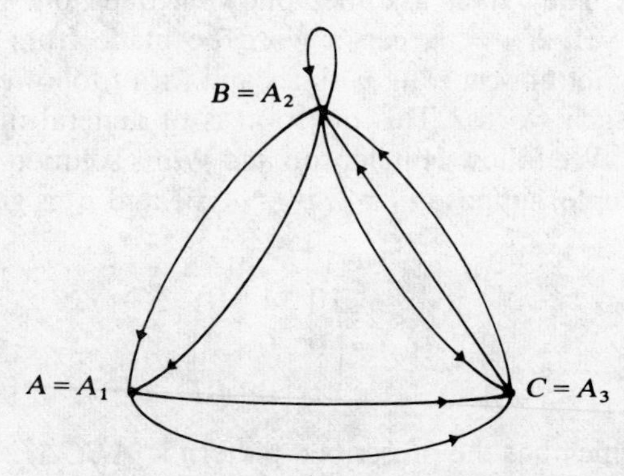

FIG. 8.9. *Multidigraph corresponding to* (8.3).

It is easy to see, using Theorem 8.4, that if conditions (8.4) and (8.5) hold for some i_1 and i_q, $i_1 \neq i_q$, and if D is weakly connected up to isolated vertices, then there is a solution word, and every solution word corresponds to an eulerian path which begins in A_{i_1} and ends in A_{i_q}. In our second example, conditions (8.4) and (8.5) hold with $i_1 = 2$, $i_q = 3$. There are a number of eulerian paths from $B = A_{i_1}$ to $C = A_{i_q}$, each giving rise to a solution word. A second example is *BACBBCBAC*.

What if the solution word begins and ends in the same letter, i.e., if $i_1 = i_q$? Then there is an eulerian closed path, and we have

$$\text{(8.6)} \qquad \sum_{k=1}^{n} v_{ik} = \sum_{k=1}^{n} v_{ki}, \quad \text{all } i.$$

Also, (8.5) holds for i_1. Condition (8.6) says that in (v_{ij}), every row sum equals its corresponding column sum. Conversely, if (8.6) holds and (8.5) holds for i_1, and D is weakly connected up to isolated vertices, then there is a solution and every solution word corresponds to an eulerian closed path in the multidigraph D, beginning and ending at A_{i_1}. This is the situation in our first example with $i_1 = 1$.

In sum, if there is a solution word, then D is weakly connected up to isolated vertices and (8.5) holds for some i_1. Moreover, either (8.6) holds, or for i_1 and some i_q, $i_1 \neq i_q$, (8.4) holds. Conversely, suppose D is weakly connected up to isolated vertices. If (8.5) holds for some i_1 and (8.4) holds for i_1 and some i_q, $i_1 \neq i_q$, then there is a solution and all solution words correspond to eulerian paths beginning at A_{i_1} and ending at A_{i_q}. If (8.5) holds for some i_1 and (8.6) holds, then there is a solution and all solution words correspond to eulerian closed paths beginning and ending at A_{i_1}.

8.6. Telecommunications. The following problem arises in telecommunications, and is discussed in Liu (1968). We follow that discussion. A rotating drum has 8 different sectors. The question is: can we tell the position of the drum without looking at it? One approach is by putting conducting material in some of the sectors and nonconducting material in others of the sectors. Place three terminals adjacent to the drum so that in any position of the drum, the terminals adjoin three consecutive sectors, as shown in Fig. 8.10. A terminal will be activated if it adjoins a sector with conducting material. If we are clever, then the pattern of conducting and nonconducting material will be so chosen that the pattern of activated and nonactivated terminals will tell us the position of the drum.

We can reformulate this as follows. Let each sector receive a 1 or a 0. We wish to arrange 8 0's and 1's around a circle so that every sequence of three consecutive digits is different. More generally, we wish to arrange 2^n 0's and 1's around a circle so that every sequence of n consecutive digits is different. Can this be done? If so, how? To see the solution, let us define a digraph D as follows. The vertices are strings of 0's and 1's of length $n-1$. There is an arc from string $a_1a_2 \cdots a_{n-1}$ to the two strings $a_2a_3 \cdots a_{n-1}0$ and $a_2a_3 \cdots a_{n-1}1$. Label each

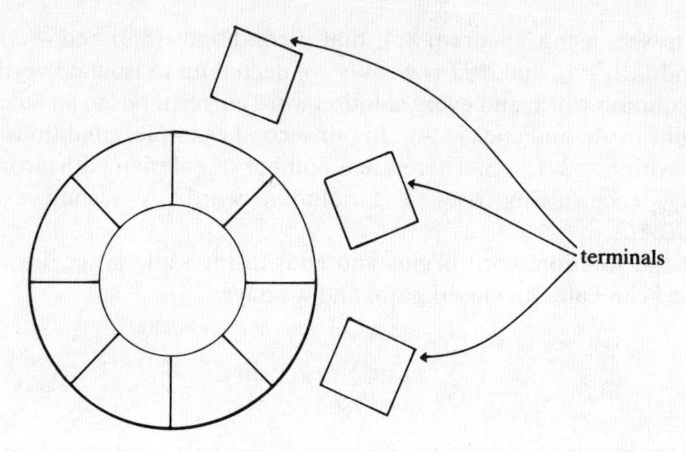

FIG. 8.10. *A rotating drum with 8 sectors and 3 adjacent terminals.*

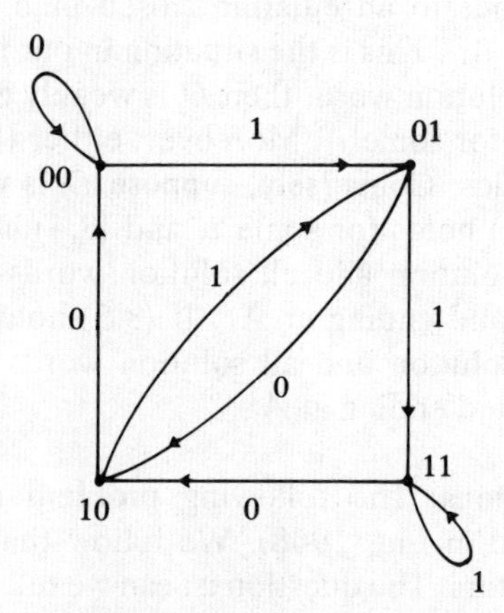

FIG. 8.11. *Digraph for solution of rotating drum problem.*

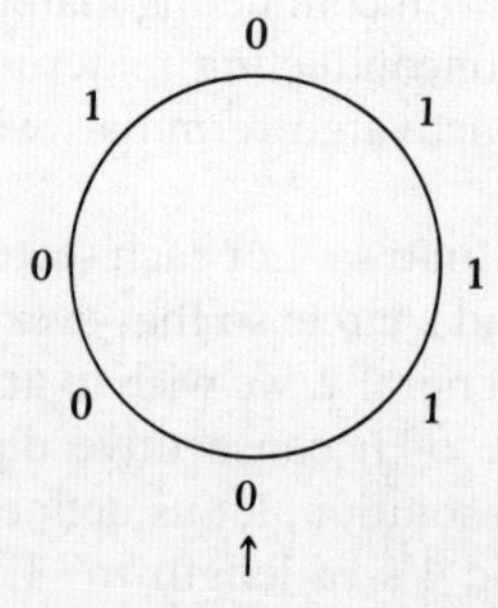

FIG. 8.12. *Arrangement of* 0's *and* 1's *which solves the rotating drum problem.*

arc with the new digit added. The resulting digraph with $n=3$ is shown in Fig. 8.11. It is easy to see that every vertex of the digraph D has indegree equal to outdegree, and hence there is an eulerian closed path. Any such path gives the desired solution, if we use the sequence of arc labels. In our example, one such eulerian closed path is 00 to 00 to 01 to 11 to 11 to 10 to 01 to 10 to 00. The corresponding arrangement of arc labels is

$$0\ 1\ 1\ 1\ 0\ 1\ 0\ 0.$$

If we arrange these around a circle as shown in Fig. 8.12, the following sequences of consecutive digits occur going counterclockwise beginning from the arrow: 011, 111, 110, 101, 010, 100, 000, 001. These are all distinct, as desired. Thus, each position of the drum can be uniquely encoded.

CHAPTER 9

Balance Theory and Social Inequalities

We shall turn in the remainder of these notes to applications of signed and weighted graphs and digraphs. A graph or digraph is *signed* if there is a sign (+ or −) on each edge or arc. It is called *weighted* if there is a real number on each edge or arc. We shall sometimes consider signed graphs or digraphs as special cases of weighted graphs or digraphs, by replacing a sign + or − by a weight +1 or −1. In a signed graph or digraph, we shall associate a sign with a chain or path. The *sign* will be + if the number of − signs on the chain or path is even, and − otherwise.

In this chapter, we shall consider some problems of interest to social scientists. One main thrust of the work is the attempt to understand social inequalities. What is it about such characteristics as sex, race, occupation, education, and so on that leads to inequalities in social interaction? We shall begin with the theory of balance, which is a forerunner for much of the work on social interaction.

9.1. The theory of balance. A great deal of work in twentieth century sociology has concerned itself with the behavior of small groups of individuals. Perhaps the simplest approach to studying such a group is to draw a digraph in which the individuals are the vertices, and in which there is an arc from vertex x to vertex y if x is in some relation to y, for example, if x likes y, x associates with y, x chooses y for a business partner, etc. Such a digraph is sometimes called a *sociogram*. Many of the relationships of interest have natural opposites, for example likes/dislikes, associates with/avoids, and so on. In that case, wc can include two different relations in one digraph by using two different kinds of arc, or by using signs to distinguish them. Then, the presence of an arc means that one of the relationships is present, and the + indicates one of the relationships, the positive one, while the − indicates the other relationship. For example, we might let an arc from x to y mean that x has strong feelings toward y, and put a + if these feelings are liking, a − if they are disliking. We obtain a signed digraph.

Let us for the sake of discussion deal with the concrete relation liking/disliking and let us assume that it is symmetric, so that we can summarize the information in a signed graph. The possible signed graphs if there are three individuals all of whom have strong feelings toward each other are shown in Fig. 9.1. Going back to the work of Heider (1946), it has been observed that groups of types I and III tend to work well together, work without tension, and so on. Groups of types II and IV do not. For example, in a group of type II, b likes both a and c, and would like to cooperate with them, but a and c dislike each other.

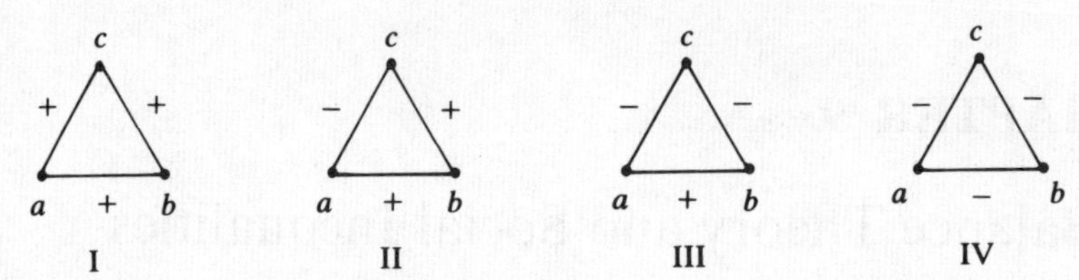

FIG. 9.1. *The possible signed graphs if there are* 3 *individuals all of whom have strong feelings toward each other.*

This causes tension. The same kind of tension does not appear in a group of type III, where a and b like each other, and both dislike c. They are perfectly content to let c work on his own, and c is quite satisfied with this arrangement.

Sociologists have used the imprecise term *balance* to describe groups which work well together, lack tension, etc. In general, groups of types I and III tend to be balanced, while groups of types II and IV do not. But what about groups whose signed graphs are more complicated? The signed graphs of types I and III can be distinguished from those of types II and IV because the former form circuits of positive sign (an even number of $-$ signs) while the latter form circuits of negative sign. This observation led Cartwright and Harary (1956) to suggest calling a signed graph and hence its underlying group of individuals *balanced* if *every* circuit was positive. In this sense, the signed graph of Fig. 9.2 is balanced, for the circuits are a, b, d, a; b, c, d, b; d, e, f, e; and a, b, c, d, a, and each of these has exactly two $-$ signs. The following theorem characterizes balanced signed graphs.

THEOREM 9.1 (Harary (1954)). *A signed graph is balanced if and only if the vertices can be partitioned into two classes so that every edge joining vertices within a class is* $+$ *and every edge joining vertices between classes is* $-$.

To illustrate this theorem, we observe that two classes which provide such a partition for the signed graph of Fig. 9.2 are $\{a, c, d, f\}$ and $\{b, e\}$. It is easy to see why the existence of such a partition implies balance. For every circuit must begin in one of the classes and end there, and so has only an even number of crossings between classes, and hence only an even number of $-$ signs. The converse is also easy to prove. One shows without loss of generality that the signed graph is connected. Then, choosing a vertex u at random, one defines one class to be all the vertices joined to u by a positive chain, and the other class to be all remaining vertices. The balance hypothesis is exactly what is needed to prove that this partition has the desired properties.

Theorem 9.1 can be considered a generalization of König's theorem (Theorem 6.1), which says that a graph is bipartite if and only if it has no odd circuits. We may obtain König's theorem from Harary's by taking all edges to be $-$.

The theory of balance suggests that groups with balanced signed graphs will exhibit lack of tension, work well together, and so on. For a discussion of tests of this theory, the reader is referred to Taylor (1970).

Balance theory has been applied to a variety of problems outside of sociology. In particular, it has been applied to the study of international relations, where the vertices become nations and the relation is allied with/allied against. It has been applied to political science, where the vertices are politicians and the relation is agrees with/disagrees with. It has been applied to the analysis of literature, where the vertices are characters in a novel, play, or short story, and the relation is liking/disliking. It is hypothesized that at a stage of tension in such a piece of work, the main characters will exhibit an unbalanced signed graph. Later, the tension will be resolved by changing a sign of a major relationship to obtain balance.

For a more detailed discussion of balance theory, with many references, see Taylor (1970). See also Roberts (1976a, § 3.1).

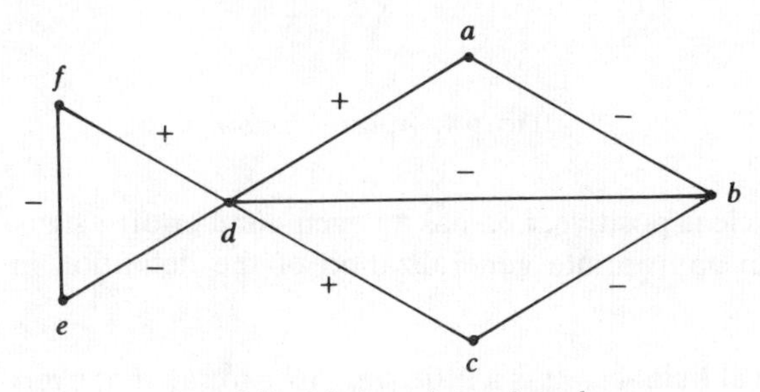

FIG. 9.2. *A balanced signed graph.*

9.2. Balance in signed digraphs. Let us briefly discuss how the notion of balance generalizes to a signed digraph. It is natural to try to call a signed digraph balanced if every cycle is positive. However, this makes the signed digraph of Fig. 9.3 vacuously balanced, while there is tension in this situation: a likes both b and c, and would like to work with them both, but c dislikes b.

An appropriate generalization of the balance concept is obtained by disregarding direction of arcs. To be precise, we say that a *semipath* in a (signed) digraph is a sequence $u_1, a_1, u_2, a_2, \cdots, u_t, a_t, u_{t+1}$, where the u_i are vertices, the a_i are arcs, and a_i is the arc (u_i, u_{i+1}) or the arc (u_{i+1}, u_i). That is, in a semipath, arcs may be traversed in either direction. The *length* of the semipath is t. In Fig. 9.4, a, (a, c), c, (d, c), d is a semipath of length 2. A *semicycle* is a semipath in which $u_1 = u_{t+1}$ and all the vertices $u_1, u_2, \cdots, u_t$ are distinct and all the arcs $a_1, a_2, \cdots, a_t$ are distinct. Thus, for example, in Fig. 9.4, a, (a, b), b, (c, b), c, (a, c), a is a semicycle. So is a, (b, a), b, (c, b), c, (a, c), a. These are different as the first arc used differs in the two cases. The *sign* of a semipath is defined analogously to the sign of a path. We say a signed digraph is *balanced* if and only

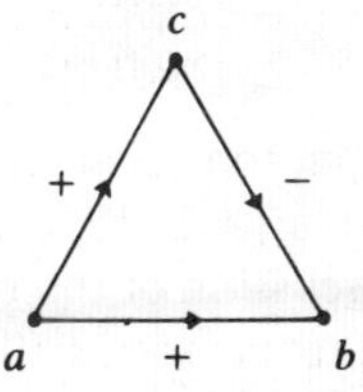

FIG. 9.3. *An unbalanced signed digraph with no negative cycles.*

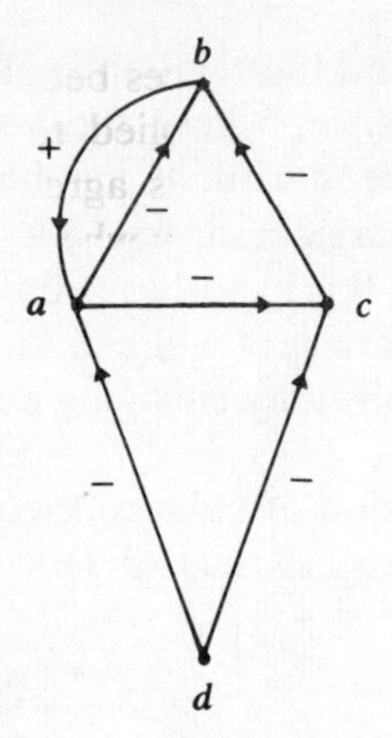

FIG. 9.4. *A signed digraph.*

if every semicycle is positive, i.e., has an even number of − signs. This definition seems to be an appropriate generalization of the definition in the symmetric case.

9.3. Degree of balance. It is a little simplistic to say that every group is either completely balanced or completely unbalanced. Rather, it probably makes more sense to speak of degrees of balance.

Let us discuss briefly some ways of measuring balance in a signed graph or digraph. One natural way to measure balance is to use the ratio p/t of the number of positive circuits (semicycles) to the total number of circuits (semicycles). An alternative is to use the ratio p/n, where n is the number of negative circuits (semicycles). Variants on this approach take account of the length of a circuit (semicycle), counting shorter circuits (semicycles) as being at least as important, and possibly more important, than longer ones. One way of taking account of length is the following. Let p_m be the number of positive circuits (semicycles) of length m, n_m the number of negative circuits (semicycles) of length m, and $t_m = p_m + n_m$. Then if $f(m)$ is a measure of the relative importance of circuits (semicycles) of length m, we might use

$$\frac{\sum_m p_m f(m)}{\sum_m t_m f(m)} \tag{9.1}$$

or

$$\frac{\sum_m p_m f(m)}{\sum_m n_m f(m)}. \tag{9.2}$$

The function $f(m)$ might be monotone nondecreasing, and might be something like $f(m) = 1/m$, $f(m) = 1/m^2$, $f(m) = 1/2^m$, etc. The function $f(m) \equiv 1$ gives the measures p/t and p/n. For a discussion of the measures p/t and p/n, see Harary (1959c). For a discussion of the measures (9.1) and (9.2), and an axiomatic derivation of them, see Norman and Roberts (1972a,b).

An alternative is to count the smallest number of signs whose negation would result in balance. This count is called the *line index for balance.* Harary (1959c) proves that in a signed graph (digraph) the line index is the same as the smallest number of edges (arcs) whose removal results in balance.

The measures (9.1) and (9.2) are equivalent in the sense that once $f(m)$ has been determined, one signed graph (digraph) is more balanced than another under (9.1) if and only if this is also the case under (9.2). However, the measure (9.1) and the line index for balance are not equivalent, i.e., there are simple examples of signed graphs (digraphs) G_1 and G_2 so that G_1 is more balanced under the line index than G_2 but less balanced under measure (9.1). For a discussion of other measures of balance, see Taylor (1970).

9.4. Distributive justice. So far, we have dealt with the situation where the vertices of a signed graph or digraph are individuals in a group. Interesting results are obtained if we allow the vertices to be other variables as well. In this section we shall use this idea to discuss the theory of distributive justice in sociology. This theory is concerned with the relation between such *characteristics* as sex, race, hair color, etc., and such *goal objects* or *rewards* as salaries, promotions, privileges, etc. The key idea is one of expectation: an individual compares his characteristics and rewards to those of others, and wants to see if justice has been done.

A theory of distributive justice is sketched out in Zelditch, Berger, and Cohen (1966) and Berger, Zelditch, Anderson and Cohen (1972). To treat the simplest case of this theory graph-theoretically, we follow Norman and Roberts (1972b). Let P and O be two individuals, and let P' be P considered as a referrent for evaluation by P. (P' can be thought of as the image P has of himself.) We study the situation from the point of view of P. We study one characteristic and one goal object, assume that the characteristic is *relevant* to the goal object, and that the characteristic and goal object can each have one of two *states*, high or low. Let GO(P) and GO(O) denote the states of the goal objects possessed by P and by O respectively. We build a signed digraph as follows. The vertices are P, P', O, GO(P), and GO(O). We draw an arc from P to GO(P) and to GO(O). On each of these arcs, we put a sign indicating the *evaluation* by P of the state, + for the high state, − for the low state. We draw arcs from P to P' and P to O. On each of these arcs, we put a sign indicating the evaluation by P, + if the individual in question (P' or O) possesses the high state of the characteristic, − if he possesses the low state of the characteristic. Finally, we draw arcs from P' to GO(P) and O to GO(O) with + signs indicating *possession.* The signed digraph obtained is shown in Fig. 9.5. For example, the characteristic might be education and the goal object salary. We feel education is relevant to salary. An individual is positively evaluated if he has a high level of education. A salary is positively evaluated if it is high.

Under the assumption that the parameter $f(m)$ is positive, we have calculated the balance measure (9.1) for each choice of sign in this digraph. The results are shown in Table 9.1 below. Cases 1, 4, 13, and 16 correspond to what Berger et al. (1972) call Justice. In each case, the state of the characteristic matches the state of the reward. In Cases 6, 7, 10, and 11, both individuals are unjustly rewarded. In Case 10, for example, P is over-rewarded (he has a low level of education and a high salary) while O is under-rewarded. This is a situation of Guilt. The remaining cases have one individual justly rewarded, but the other not. The balance measure indicates that the situations of Justice are perfectly balanced, the situations where both individuals are unjustly rewarded are perfectly unbalanced, and the situations where one individual is unjustly rewarded are in between. By adding more information to this signed digraph, Norman and Roberts (1972b) show how further distinctions among the different cases can be made.

For example, we can study coalition formation to distinguish the four cases where both individuals are unjustly rewarded. We draw an arc from P' to O representing attraction/repulsion and ask whether the arc should receive a + or − sign, + for attraction, − for repulsion. We obtain the signed digraph of Fig. 9.6. We have calculated the numerator of the balance measure (9.1) for the two choices of sign on this arc in each of the cases 6, 7, 10, and 11. (The denominator is always the same in all cases, since all have the same semicycles.) The results are shown in Table 9.2. Suppose we assume that P' and O form a coalition if and only if the signed digraph is more balanced with attraction than with repulsion. Now coalition formation occurs in Cases 6 and 11 if and only if

$$f(3)+f(5)>2f(4)$$

and in Cases 7 and 10 if and only if

$$2f(4)>f(3)+f(5).$$

Hence, we conclude that either there is no coalition formation ($f(3)+f(5)=2f(4)$), or it takes place only in situations where both individuals are treated with equal injustice (Cases 6 and 11), or it takes place only where the two individuals are treated with opposite injustice (Cases 7 and 10). Notice that this conclusion follows independently of the *specific value* of the parameter f. Adding more

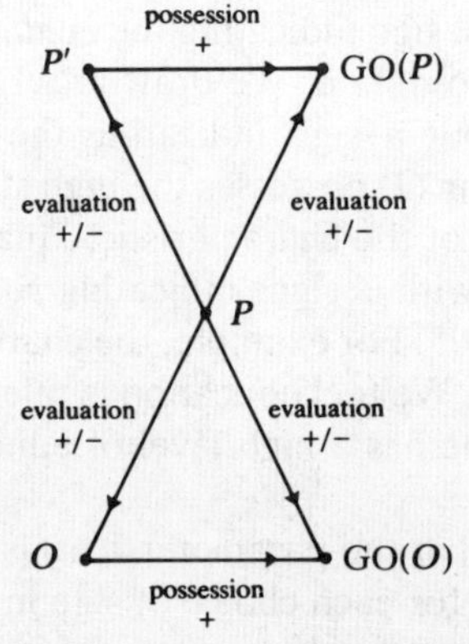

FIG. 9.5. *Signed digraph for the distributive justice situation.*

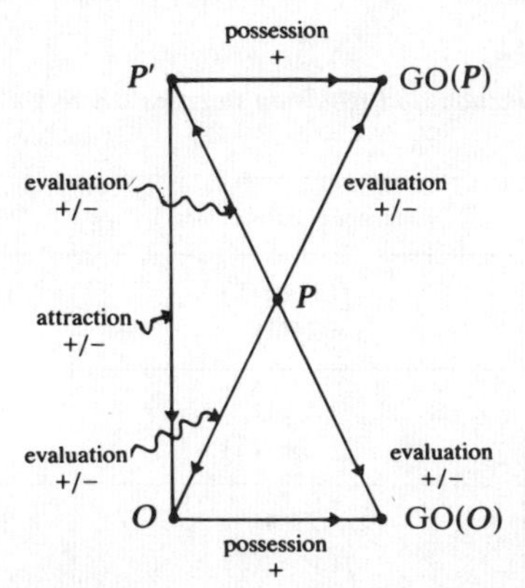

FIG. 9.6. *Signed digraph for the distributive justice situation with attraction/repulsion between P′ and O added.*

information to the digraph allows the conclusion that coalition formation will take place in situations of equal injustice. We shall not go into further detail here. Rather, we simply mentioned this example to illustrate the kind of analysis which is sometimes made using signed digraphs where the vertices are other than just individuals and where the arcs represent different kinds of relationships.

TABLE 9.1
Relative balance in 16 cases of signed digraph of Fig. 9.5.

Case Number	*PP′*	*PGO(P)*	*PO*	*PGO(O)*	Balance	Description
1	+	+	+	+	1	Justice
2	+	+	+	−	$\frac{1}{2}$	Other unjustly rewarded (under-reward)
3	+	+	−	+	$\frac{1}{2}$	Other unjustly rewarded (over-reward)
4	+	+	−	−	1	Justice
5	+	−	+	+	$\frac{1}{2}$	Self unjustly rewarded (under-reward)
6	+	−	+	−	0	Both unjustly rewarded
7	+	−	−	+	0	Both unjustly rewarded
8	+	−	−	−	$\frac{1}{2}$	Self unjustly rewarded (under-reward)
9	−	+	+	+	$\frac{1}{2}$	Self unjustly rewarded (over-reward)
10	−	+	+	−	0	Both unjustly rewarded
11	−	+	−	+	0	Both unjustly rewarded
12	−	+	−	−	$\frac{1}{2}$	Self unjustly rewarded (over-reward)
13	−	−	+	+	1	Justice
14	−	−	+	−	$\frac{1}{2}$	Other unjustly rewarded (under-reward)
15	−	−	−	+	$\frac{1}{2}$	Other unjustly rewarded (over-reward)
16	−	−	−	−	1	Justice

TABLE 9.2

Relative balance of attraction and repulsion in cases where both individuals are unjustly rewarded.

Case Number	Numerator of Balance Measure	Description
6, with +	$f(3)+f(5)$	Both under-rewarded
6, with −	$2f(4)$	
7, with +	$2f(4)$	Self under, other over
7, with −	$f(3)+f(5)$	
10, with +	$2f(4)$	Self over, other under
10, with −	$f(3)+f(5)$	
11, with +	$f(3)+f(5)$	Both over-rewarded
11, with −	$2f(4)$	

9.5. Status organizing processes and social inequalities. Since the early twentieth century, sociologists have been concerned with *status-organizing processes*, processes by which differences in evaluations and expectations of individuals affect social interactions. A major goal of studying such processes is the explanation of social inequalities. The theory of distributive justice was an early theory in the work of Berger and his colleagues, which tried to explain the tension resulting from violation of expectations by showing that certain signed digraphs were unbalanced, or relatively unbalanced. In Berger et al. (1977), chains in a signed graph are used to study induced expectations. The basic idea is that the variables are individuals, goal objects, states, and so on, much as in our discussion of § 9.4, except that direction of relationships is disregarded. In the resulting signed graphs, if there is a chain from an individual P to a goal object G, the sign of this chain indicates whether the individual is expected to be rewarded with the high or the low state of the goal object. If there are several chains from P to G of the same sign, this leads to stronger expectations, while if there are chains of opposite sign, this leads to inconsistent expectations (there is imbalance). The expectations are calculated quantitatively using a parameter much like $f(m)$, and are used to predict reactions of one individual to another when they disagree.

9.6. Strengths of likes and dislikes. One of the weaknesses of the signed graph approach to balance is that it omits strengths of likes and dislikes. One approach which attempts to use such strengths is due to Hubbell, Johnson, and Marcus (1978). Let the individuals in a group be labeled as $1, 2, \cdots, n$. Let $s_{ij}^{(t)}$ be the sentiment of i for j at time t. How do sentiments change over time? Hubbell et al. assume changes take place only at discrete times $t = 0, 1, 2, \cdots$ and make the following assumption. The sentiment s_{ik} of i for k can be changed by sentiments of i for j and j for k—this is called an *inductive change*—and by sentiments of i for j and k for j—this is called a *comparative* *change.* Assuming these effects take place equally, Hubbell et al. specify that

$$s_{ik}^{(t+1)} = \frac{1}{2}\left(\sum_j s_{ij}s_{jk} + \sum_j s_{ij}s_{kj}\right). \tag{9.3}$$

In terms of matrices, if $S_t = (s_{ij}^{(t)})$, then (9.3) becomes

$$S_{t+1} = \frac{1}{2}(S_t^2 + S_t S_t^T) \tag{9.4}$$

where S_t^{T} is the transpose of S_t.

Given a matrix S, we can associate a sign pattern with it. For example, the matrix

$$S = \begin{pmatrix} 75 & -6 & 0 \\ -1 & 7 & 0 \\ 0 & 0 & 0 \end{pmatrix} \tag{9.5}$$

has the sign pattern

$$\begin{pmatrix} + & - & 0 \\ - & + & 0 \\ 0 & 0 & 0 \end{pmatrix}. \tag{9.6}$$

The sign pattern defines a signed digraph in a natural way. Under the assumption (9.4), the sequence of matrices might or might not approach a limit. (Hubbell et al. discuss this problem matrix-theoretically.) Of more interest to us is the corresponding sequence of sign patterns. Does this approach a limit, i.e., is there some point beyond which all of the sign patterns are the same? If so, does this sign pattern correspond to a balanced signed digraph? How can we tell if a given matrix $S = S_1$ gives rise to a sequence of sign patterns which eventually stabilizes? How can we tell if the sequence eventually stabilizes to a pattern corresponding to a balanced signed digraph? Does the answer depend on the specific numbers in S?

We have asked lots of questions. Let us give an example. The matrix S of (9.5) has the property that if $S = S_1$, then all matrices S_t defined by (9.4) have the same sign pattern (9.6). The corresponding signed digraph is balanced. Thus, we know that if the initial sentiments of the individuals are given by (9.5), and if they change according to the Hubbell et al. model (9.4), then the group will always be balanced. The conclusion depends only on the sign pattern of S, and not on the specific entries. The advantage of the model we have discussed over those using only signs is that in the latter, changes of sign can take place only abruptly, while here sentiments can change gradually.

The questions we have raised (which are also raised by Hubbell et al.) seem to be difficult ones, and to our knowledge, no one has made progress on them.

However, we shall see in the next two chapters that one can answer a large number of similar questions which arise in public policy situations, economic analysis, applied ecology, and the like. These questions deal with matrices whose elements are known only up to sign. The questions then ask: can we draw various conclusions about such things as systems of linear equations or systems of differential equations which use these matrices for coefficients, if we know the entries in the matrices only up to sign? We shall see that we can draw many such conclusions, by making use of the signed digraph corresponding to the sign pattern.

CHAPTER 10

Pulse Processes and their Applications

10.1. Structural modeling. Many mathematical approaches to the study of complex systems emphasize a careful study of their numerical properties. These often end up with an optimization problem, or otherwise deal with matters which are closely tied to specific numbers. Kane (Kane (1973), Kane et al. (1973)) calls such approaches *arithmetic*, in contrast to the approaches which emphasize structure, pattern, shape, or general tendency, with little emphasis on specific values. The latter approaches Kane calls *geometric*. The very complexity of real-world systems involving energy, food, transportation, communications, and the like makes it very difficult to get precise enough information to approach them arithmetically. This has given rise to an increasingly widespread effort to develop techniques for studying systems geometrically. One general geometric approach, which relates geometric properties of complex systems to certain structural properties of these systems, is coming to be known as *structural modeling*. Surveys of structural modeling techniques have recently been carried out by a number of authors. For information, see Cearlock (1977) and Lendaris and Wakeland (1977).

Most structural modeling techniques start out by identifying variables relevant to a problem being studied. It is usually assumed that more complicated interrelations among these variables are captured by their pairwise interactions. A digraph is constructed with vertices the variables and with an arc from variable x to variable y if x is in some strong relation to y, commonly the relation "causes" or "effects." Loops are allowed. Usually, as in our discussion of the previous chapter, the relation in question is treated as "signed," and the sign of the effect of x on y is identified. Thus, a signed digraph is constructed. Beyond that, strength of effect is sometimes taken into account, and a real number or weight measuring the strength of the effect of x on y is obtained. The resulting digraph with a real number $w(x, y)$ on each arc (x, y) is called a *weighted digraph*, and sometimes a *structural model*. The exact procedures for constructing the structural model, for obtaining and interpreting the arcs, signs, or weights, and for analyzing the structural model, vary from methodology to methodology. We shall present just one such approach in the following sections, and apply it to the relationship between energy use and food production. The method we discuss has been applied to energy use, air pollution, and transportation systems (Roberts (1971c), (1973), (1974), (1975), (1976a), (1976b), Roberts and Brown (1975)). Other structural modeling approaches have been applied to the analysis of coastal resources (Coady et al. (1973)), health care delivery in British Columbia (Kane, Thompson and Vertinsky (1972)), Naval manpower (Kruzic (1973a)), Canadian water policy (Kane, Vertinsky and Thompson (1972), Kane

(1973)), Canadian environmental policy (Kane (1973)), transportation problems (Kane (1972)), U.S. energy policy (Kruzic (1973b)), the use of coal in inland waterways (Antle and Johnson (1973)), the study of ecosystems (Levins (1974a,b)), the deliberations of governmental bodies and historical personages (Axelrod (1976)), and so on. Variants of the method we discuss have been developed by McLean (1976) and McLean and Shepherd (1976), (1978a,b).

10.2. Energy and food. As world population continues to grow, there is increasing concern about mankind's ability to feed itself. Although we have begun to see famine in parts of the world, more serious starvation has been avoided so far by the tremendous productivity of modern agricultural systems. Unfortunately, this productivity has come at a price: the use of vast amounts of energy. Energy is used in modern agriculture in the form of fossil fuel to run farm machines, in irrigation, in the production and application of chemical fertilizers and pesticides, and so on. Yields of food on a particular field have risen dramatically; for example, Pimentel et al. (1973) and Pimentel et al. (1974) report that yields of corn increased from about 85 bushels per hectare in 1945 to about 203 in 1970, or almost two and one half times as much. However, during the same time span, total energy input into a one hectare corn field essentially tripled. With the increasing scarcity of energy, it might not be possible to continue the pace of production of modern agriculture, let alone to expand this pace.

For the sake of discussion, we construct a small talking-purpose example of a structural model to discuss this problem. We shall consider world food production under U.S. style agriculture (a highly productive, highly energy-intensive kind of agriculture). We assume that population is not curtailed, and continues to expand as food becomes available. We shall use five variables in our simple example. These are population (P), demand for food (D), food yield (Y), cost of food (R), and energy input into food production (E). The units we shall use are identified in Table 10.1 and the current (as of 1970) values of these variables are estimated there.

TABLE 10.1

Variable	Unit	Current Value*
Population (P)	billions of people	4
Demand for food (D)	10^{15} kcal (kilocalories) per year	4
Food yield (Y)	millions of kcal/hectare/ year	17.9
Cost of food (R)	dollars per person per year	\$700
Energy input into food production (E)	millions of kcal/hectare/ year	7.1

* The values of Y and E are based on data for corn production in the U.S. and are obtained from Pimentel et al. (1973) or Pimentel et al. (1974). Cost of food is estimated from U.S. figures obtained from an almanac. Demand for food is calculated under the assumption that one person requires approximately 3000 kcal per day and there are 4 billion people.

We shall use these variables as vertices of a signed digraph. We draw an arc from variable x to variable y if a change in x has a significant effect on y. Although loops are allowed, we choose to treat them as insignificant in our example. We put a + sign on the arc (x, y) if, all other things being equal, a change in x augments y, and a − sign if, all other things being equal, the change inhibits y. (The effect is said to be *augmenting* if whatever happens at x is reflected by a change in the same direction at y, i.e., increase leads to increase, decrease to decrease, and the effect is said to be *inhibiting* if whatever happens at x is reversed in y, i.e., increase leads to decrease and decrease to increase.) Figure 10.1 shows such a signed digraph obtained from the five variables we have identified. Note that one could argue that other arcs should be included, but we have included only those which seem most important. In particular, we have chosen to disregard the effect of demand on price, but to include the inhibiting effect of price on demand. In this structural model, supply (yield) and energy use are the primary direct influences on price.

Certain things can be learned about the food-energy system without knowing any more information about it than what is contained in a signed digraph like that of Fig. 10.1. For example, we note that there are three cycles in this digraph, D, E, R, D; P, D, E, Y, P; and D, E, Y, R, D. The latter two are positive cycles (the number of − signs is even), while the former is negative. Cycles correspond to feedback processes, and positive cycles, it is easy to see, to positive feedback. Hence, we see two important processes which lead to positive feedback, and tend to put pressures on the system to continue changing in the direction of an initial change. One such process starts with increasing population, which leads to more demand for food, which leads to increased use of energy in food production, which leads to more food yield, which makes it possible for population to grow further. We also see one important negative feedback process, that operating through the increase in price of food when more energy is used in food production.

The signed digraph makes certain omissions and certain simplifications. The simplifications include the tacit assumption that two variables are always related in the same way (in an augmenting or inhibiting relationship), independent of their levels. The omissions include no discussion of strength of effect and no discussion of the time it takes for an effect to take place. We shall discuss the addition of strength of effect, but not the time lag problem.

We can include strengths of effects by adding weights to a signed digraph. The weight $w(x, y)$ on the arc (x, y) will have in our approach the following specific interpretation: whenever variable x increases by u units, then variable y increases by $u \cdot w(x, y)$ units. This is a very special linear assumption, and most variables are not related in quite so simple a way. However, we shall use this assumption to gather weights, understanding that the use of the same weight independent of time and of the levels of the variables in question is a rather serious simplifying assumption. Weights with the stated interpretation have been obtained for the signed digraph in Fig. 10.1. They are included as weights in a weighted digraph in Fig. 10.2. We shall not discuss how these weights were obtained, except to say that the procedure was, as is usually the case, a subjective one, and involved some use of real data (for example that of Pimentel et al. (1973) and Pimentel et al. (1974)), and some guesswork. Note that, for example, the weight -1.4 on the arc (Y, R) means that every time food yield increases by 1 unit, i.e., 1 million kcal/hectare/year, then food cost goes down by \$1.40 per person per year. This is assumed true independent of the level of yield or the current price—undoubtedly a somewhat faulty assumption.

Under the stated interpretation of weights, it makes sense to associate a weight with a cycle as well as with an arc, and to do so by multiplying the weights of the arcs on that cycle. If this is done, we note that the cycle P, D, E, Y, P has weight 1.613 and the cycle D, E, Y, R, D has weight .025. This suggests that the former cycle will correspond to increasingly larger augmentations (since its weight is greater than 1), while the latter cycle will correspond to increasingly smaller augmentations. In the next few sections, we shall analyze the use of the weighted digraph to predict future levels of the different variables, and we shall ask to what extent the corresponding signed digraph can help us in making such predictions.

Even without further analysis, the signed and weighted digraphs constructed have given us certain insights into the food-energy system. Indeed, the very

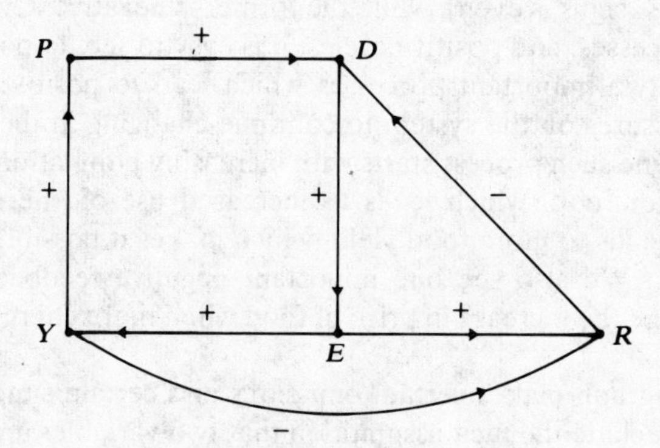

FIG. 10.1. *Signed digraph for the food-energy system.*

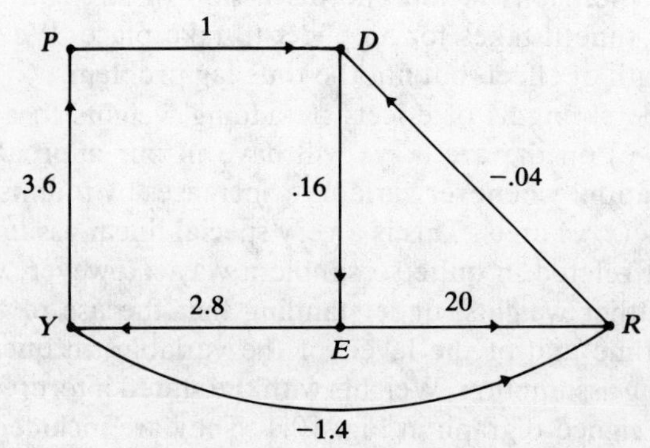

FIG. 10.2. *Weighted digraph for the food-energy system.*

process of constructing them has forced us to concentrate on the basic variables in question and on the fundamental interrelations among these variables. Analysis of these digraphs has helped us to identify basic sources of feedback, and has helped us to begin to understand the dynamic relationships among the variables in question. In this sense, the tools of graph theory are to be regarded as learning devices, used to help us understand complex systems. The more quantitative models we shall discuss in the next few sections also play primarily the role of a learning device, in helping us to identify basic processes, understand basic interrelationships, and pinpoint possible strategies to analyze further.

10.3. Pulse processes. Let us suppose D is a weighted digraph and $u_1, u_2, \cdots, u_n$ are its vertices. We shall assume that each vertex u_i attains a *value* or level $v_i(t)$ at each time t, and that time takes on discrete values, $t=0, 1, 2, \cdots$. We shall be interested in the *change in value* or the *pulse* at vertex u_i at time t. This will be denoted $p_i(t)$ and will be obtained by

$$p_i(t)=v_i(t)-v_i(t-1) \quad \text{if } t>0. \tag{10.1}$$

For $t=0$, $p_i(t)$ and $v_i(t)$ must be given as initial conditions. Then, our interpretation of the weight $w(x, y)$ gives rise to the following rule for change of values:

$$v_i(t+1)=v_i(t)+\sum_j w(u_j, u_i)p_j(t). \tag{10.2}$$

It follows from (10.1) and (10.2) that

$$p_i(t+1)=\sum_j w(u_j, u_i)p_j(t). \tag{10.3}$$

We shall say that the initial conditions $p_i(0)$ and $v_i(0)$ and the rule (10.2) define an *autonomous pulse process*. For example, in the weighted digraph of Fig. 10.2, suppose $v_i(0)=0$, all i, and $p_1(0)=1$, $p_2(0)=2$, $p_3(0)=5$, $p_4(0)=-3$, and $p_5(0)=-2$, where $P=u_1$, $D=u_2$, $Y=u_3$, $R=u_4$, and $E=u_5$. Then, for example, using (10.3), we obtain

$$p_4(1)=(-1.4)(5)+(20)(-2)=-47.$$

That is, at time 1, variable 4 decreases by 47 units. Hence,

$$v_4(1)=v_4(0)+p_4(1)=-47.$$

In an autonomous pulse process, we make some initial changes, and trace them out over time. It is easy enough to *forecast* future values $v_i(t)$. Suppose we let

$$P(t)=(p_1(t), p_2(t), \cdots, p_n(t))$$

and

$$V(t)=(v_1(t), v_2(t), \cdots, v_n(t)).$$

Let A be the *weighted adjacency matrix* of the weighted digraph D, the $n \times n$ matrix with i, j entry equal to $w(u_i, u_j)$ if there is an arc from u_i to u_j and equal to 0 otherwise. Equation (10.3) then says that

$$P(t+1) = P(t)A,$$

and hence we see that

$$P(t) = P(0)A^t. \tag{10.4}$$

It follows that

$$V(t) = V(0) + P(0)(A + A^2 + \cdots + A^t). \tag{10.5}$$

From (10.5), forecasts of future values can be made.

We shall be interested not so much in the specific values which come out of such forecasts, but, in the spirit of structural modeling, in the general question of the behavior of the sequences of values or pulses. In particular, we say that a vertex u_i is *pulse stable* under an autonomous pulse process if the sequence

$$\{|p_i(t)| : t = 0, 1, 2, \cdots\}$$

is bounded, and u_i is *value stable* under the pulse process if the sequence

$$\{|v_i(t)| : t = 0, 1, 2, \cdots\}$$

is bounded. It is easy to show from (10.1) that value stability implies pulse stability. However, the converse is false. We say a *weighted digraph* is *pulse* or *value stable* under an autonomous pulse process if every vertex of the digraph is. If a digraph is pulse or value stable under *all* autonomous pulse processes, then its structure is such that the system is not liable to "blow up" after it is initially perturbed, no matter what the nature of the initial perturbation.

The following theorems about pulse and value stability follow from results in Roberts and Brown (1975).

THEOREM 10.1. *Suppose D is a weighted digraph and A is its weighted adjacency matrix. If D is pulse stable under all autonomous pulse processes, then every eigenvalue of A has magnitude less than or equal to one.*

THEOREM 10.2. *Suppose D is a weighted digraph and A is its weighted adjacency matrix. If every eigenvalue of A has magnitude less than one, then D is pulse stable under all autonomous pulse processes.*[17]

THEOREM 10.3. *Suppose D is a weighted digraph and A is its weighted adjacency matrix. Then D is value stable under all autonomous pulse processes if and only if D is pulse stable under all autonomous pulse processes and* 1 *is not an eigenvalue of D.*

To illustrate these theorems, we note that for the weighted digraph of Fig. 10.2, the characteristic polynomial is given by

$$C(\lambda) = \lambda(\lambda^4 + .128\lambda - 1.638). \tag{10.6}$$

[17] Conditions both necessary and sufficient for pulse stability under all autonomous pulse processes are also given in Roberts and Brown (1975).

We shall see below why this is true. In any case, since the product of the roots of the polynomial in parentheses in (10.6) is plus or minus the constant term -1.638, we observe that there must be a root of magnitude larger than one. Hence, Theorem 10.1 implies that the system (weighted digraph) is pulse unstable under some autonomous pulse process and Theorem 10.3 implies that the system will therefore be value unstable under some such process. The interpretation of this result is as follows. There is some way of introducing a perturbation or change in the energy-food system which will lead to both increasing changes and values elsewhere, for example in population, energy use, or price of food. It is easy enough to see that most changes in this particular system will have this property, and in particular that any initial increase in population will. For other applications of Theorems 10.1 to 10.3, see Roberts (1976a, Chap. 4).

10.4. Structure and stability. A major problem with the eigenvalue theorems stated in the previous section is that they do not explain why instabilities occur. In this section, we shall discuss methods of determining the source of instabilities from the structure of the digraph. Once one has such methods, they can be used to find ways of changing the system which will avoid instabilities.

In particular, the kinds of results we shall describe relate the eigenvalues and characteristic polynomial of a weighted digraph (of its weighted adjacency matrix) to the structural properties of the digraphs. An early result along these lines is due to Harary. In § 1.4 we defined a strong component of a digraph to be an equivalence class of vertices under the equivalence relation where two vertices are equivalent if and only if each is reachable from the other by a path. It is easy to see that a strong component corresponds to a maximal, strongly connected, generated subgraph. The strong components partition the vertices of a digraph. Figure 10.3 shows a weighted digraph and its strong components, and

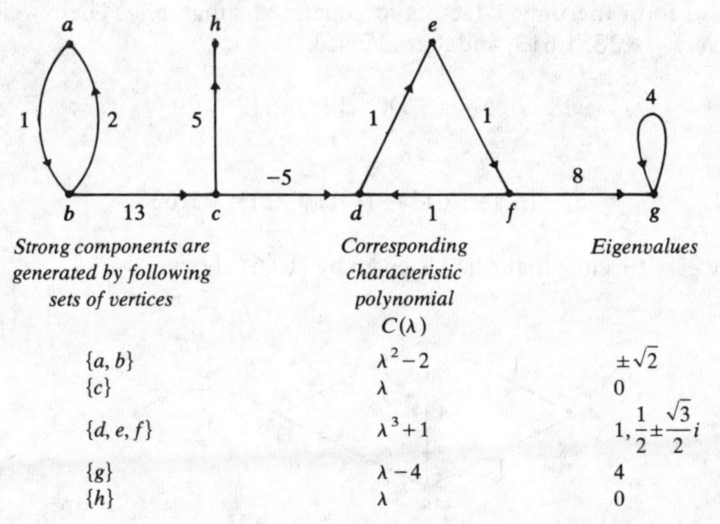

Strong components are generated by following sets of vertices	*Corresponding characteristic polynomial* $C(\lambda)$	*Eigenvalues*
$\{a, b\}$	λ^2-2	$\pm\sqrt{2}$
$\{c\}$	λ	0
$\{d, e, f\}$	λ^3+1	$1, \frac{1}{2}\pm\frac{\sqrt{3}}{2}i$
$\{g\}$	$\lambda-4$	4
$\{h\}$	λ	0

FIG. 10.3. *Use of strong components to calculate eigenvalues.*

the computation there illustrates the following results. (These results can also be turned around and used to calculate eigenvalues of an arbitrary matrix.)

THEOREM 10.4 (Harary (1959a)). *The characteristic polynomial of a weighted digraph is the product of the characteristic polynomials of its strong components.*

COROLLARY. *The set of eigenvalues of a weighted digraph is the union (counting multiplicity) of the sets of eigenvalues of its strong components.*

To state a second result along these lines, we define a digraph to be a 1-*factor* if every vertex has indegree one and outdegree one. It is easy to show that a 1-factor is a union of disjoint cycles. A 1-factor in a digraph is a spanning subgraph which is a 1-factor. The weight $w(H)$ of a 1-factor H in a weighted digraph D is the product of the weights of the arcs of H. For example, in the weighted digraph of Fig. 10.4, the cycles a, b, c, a and d, e, d define a 1-factor. Its weight is $(1)(-5)(-1)(2)(1)=10$.

THEOREM 10.5 (Chen (1967), (1971)). *Suppose D is a weighted digraph and*

$$C(\lambda)=\lambda^n+\sum_{k=1}^{n} b_k\lambda^{n-k}$$

is its characteristic polynomial. Then

$$b_k=\left[\sum_{G_k}\sum_{u}(-1)^{L_H}w(H_u)\right], \tag{10.7}$$

where G_k is a k-vertex generated subgraph of D, H_u is the u-th l-factor in G_k, L_H is the number of cycles in H_u, and $w(H_u)$ is the weight of H_u.

Let us apply this theorem to the weighted digraph of Fig. 10.2. We already observed that there are only three cycles, D, E, R, D; P, D, E, Y, P; and D, E, Y, R, D. Since each pair of these cycles has a common vertex, these cycles also form the only 1-factors in generated subgraphs. Their weights are respectively $-.128$, 1.613, and $.025$. Hence,

$$b_3=(-1)(-.128)=.128$$

and

$$b_4=(-1)(1.613)+(-1)(.025)=-1.638.$$

It is now easy to show that $C(\lambda)$ is given by (10.6) above.

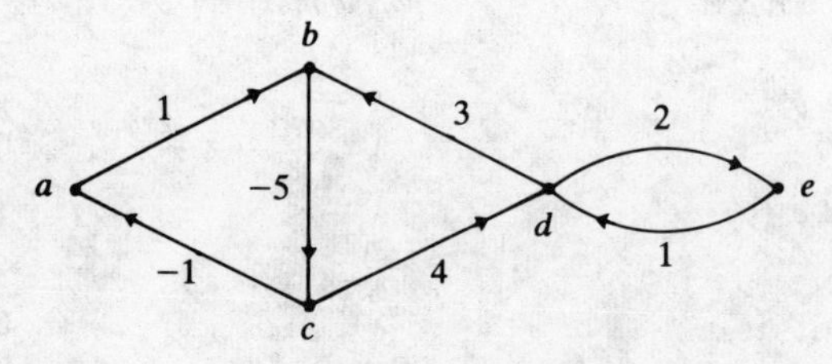

FIG. 10.4. *A* 1-*factor consists of the cycles a, b, c, a and d, e, d.*

The weighted digraph of Fig. 10.5 presents a slightly more difficult example. Using the identification of 1-factors shown in this figure, we have

$$b_1=(-1)(1)=-1,$$
$$b_2=(-1)(-5)+(-1)(1)=4,$$
$$b_3=(-1)(60)+(-1)(-2)+(-1)^2(-5)=-63,$$
$$b_4=(-1)^2(60)+(-1)^2(-2)+(-1)^2(-5)=53,$$
$$b_5=(-1)^2(10)+(-1)^2(60)=70,$$
$$b_6=(-1)^3(10)+(-1)^2(-120)=-130,$$
$$b_7=(-1)^3(-120)=120,$$
$$b_8=0,$$
$$b_9=0.$$

Hence,

$$C(\lambda)=\lambda^9-\lambda^8+4\lambda^7-63\lambda^6+53\lambda^5+70\lambda^4-130\lambda^3+120\lambda^2.$$

Theorem 10.5 gives us some hope of relating instability to structure. For, if s is the highest number so that the coefficient b_s is nonzero, then $C(\lambda)$ factors as

$$\lambda^{n-s}(\lambda^s+\cdots+b_s),$$

and hence b_s is plus or minus the product of the nonzero eigenvalues. By Theorem 10.1, the only hope for pulse stability is to have $|b_s|\leqq 1$. We see in our example of Fig. 10.2 that s is 4 and $b_s=-1.638$. Hence, we know immediately that the system will be pulse unstable under some autonomous pulse process. We also observe that the cycle with highest (absolute) weight is P, D, E, Y, P. If we could somehow break up this cycle, we might drive b_s down below 1 in magnitude, and have a chance at having created a pulse stable system. We can accomplish this in a variety of ways. In particular, eliminating the arc (Y, P) eliminates this cycle. This change corresponds to eliminating the assumption that population will be allowed to increase to the available food supply. It corresponds to some form of population control. If this arc (Y, P) is deleted in the weighted digraph of Fig. 10.2, then, using Theorem 10.5, one obtains the following characteristic polynomial:

$$C(\lambda)=\lambda(\lambda^4+.128\lambda-.025).$$

It is easy enough to show that all of the roots of this polynomial have magnitude less than one.[18] Hence, by Theorem 10.2, the new weighted digraph is pulse stable under all autonomous pulse processes. Since 1 is not a root, Theorem 10.3 implies that it is also value stable. The control on population has cut down on unbounded increases in changes or in values in the system.

[18] The roots are approximately 0, .19, $-.56$, $.19\pm.46i$.

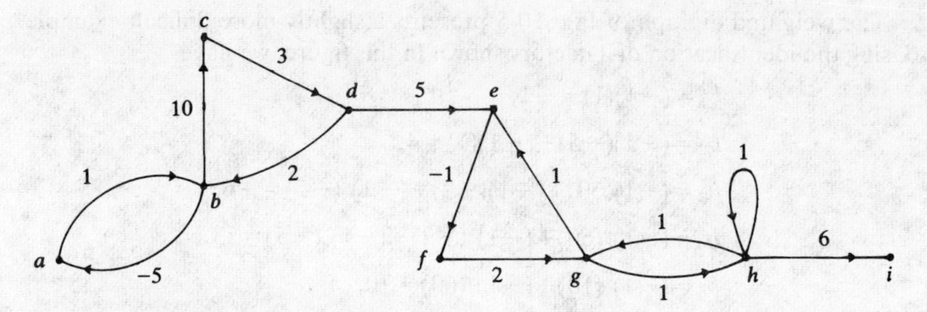

1-*factor in a generated subgraph consists of the cycles*	*number of vertices*	*weight*
h, h	1	1
a, b, a	2	−5
g, h, g	2	1
b, c, d, b	3	60
e, f, g, e	3	−2
a, b, a; h, h	3	−5
b, c, d, b; h, h	4	60
e, f, g, e; h, h	4	−2
a, b, a; g, h, g	4	−5
a, b, a; e, f, g, e	5	10
b, c, d, b; g, h, g	5	60
a, b, a; e, f, g, e; h, h	6	10
b, c, d, b; e, f, g, e	6	−120
b, c, d, b; e, f, g, e; h, h	7	−120

FIG. 10.5. *A weighted digraph and its 1-factors.*

10.5. Integer weights. Theorem 10.5 leads to some interesting results about stability if we are willing to assume that all the weights of a weighted digraph are integers. In this case, we say the digraph is *integer-weighted.*

THEOREM 10.6. *Suppose D is an integer-weighted digraph and*

$$C(\lambda)=\lambda^n+\sum_{k=1}^{n} b_k\lambda^{n-k}$$

is its characteristic polynomial. Suppose some $b_k \neq 0$ and let s be the highest subscript so that $b_s \neq 0$. If D is pulse stable under all autonomous processes, then

$$b_s = \pm 1$$

and

$$b_i = b_s b_{s-i}, \qquad i=1, 2, \cdots, s-1.$$

THEOREM 10.7. *Under the hypotheses of Theorem 10.6, D is value stable under all autonomous pulse processes if and only if D is pulse stable under all autonomous pulse processes and $\sum_{i=1}^{s} b_i \neq -1$.*

To illustrate these theorems, consider the integer-weighted digraph of Fig. 10.6. Using Theorem 10.5 we have $b_2=-5$, $b_3=-1$ and $b_5=1$. Since $s=5$ and

$b_s = 1$, the first condition of Theorem 10.6 is satisfied. However, $b_2 \neq b_5 b_3$, and so the second condition is violated, and we conclude that the digraph is pulse unstable under some autonomous pulse process. Theorems 10.6 and 10.7 are proved in exactly the same way that Theorems 7 and 8 of Roberts and Brown (1975) are proved in the special case called in that paper advanced rosettes.[19] One simply substitutes our b_k for $-a_k$ as used by Roberts and Brown.

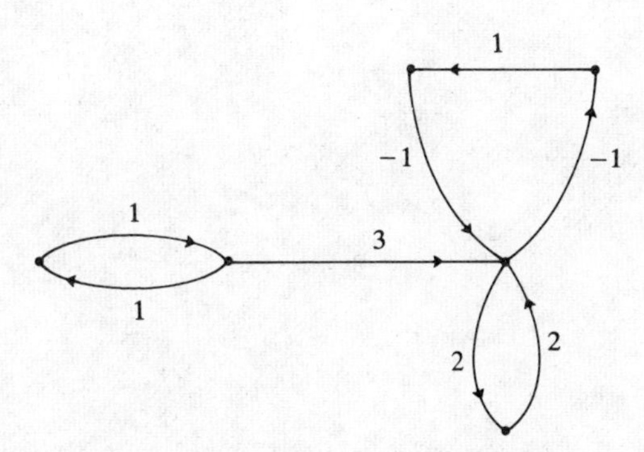

FIG. 10.6. *An integer-weighted digraph.*

10.6. Stability and signs. At the end of Chapter 9, we asked a general question, namely, what can one learn about properties of a system knowing only signs? As far as stability is concerned, the answer is very little. For, suppose D is a weighted digraph and D' is the corresponding signed digraph, i.e., a signed digraph with the same set of vertices and arcs, but signs replacing weights. Let D' be considered a weighted digraph with weights $+1$ and -1. Can we infer from the pulse or value stability or instability of D' any results about the stability of D? The answer is: almost never. For, given a signed digraph which has at least one cycle, it is easy to find an assignment of weights having the given signs for which the highest nonzero coefficient b_s in Theorem 10.5 is greater than 1. Hence, the product of the nonzero eigenvalues is larger than 1, and Theorem 10.1 implies the corresponding weighted digraph is pulse and value unstable under some autonomous pulse process. On the other hand, it is easy to show that there is always some assignment of weights having the given signs under which the resulting weighted digraph is pulse and value stable. For, let A be any real matrix. If α is any real number, the eigenvalues of αA are α-multiples of the eigenvalues of A. Hence, α can be picked small enough that all eigenvalues of αA have magnitude less than 1, and hence, by Theorems 10.2 and 10.3, αA gives rise to a pulse and value stable weighted digraph for sufficiently small α.

In the next chapter, we shall see that, in contrast to these negative results, we can sometimes learn a great deal from the sign pattern.

[19] The author thanks Professor Philip Straffin for suggesting these ideas.

CHAPTER 11

Qualitative Matrices

In this chapter, we shall study properties of a matrix or, equivalently, of a weighted digraph, which depend only on the sign of the entries or weights. In general, if D is a weighted digraph (with loops allowed), and A is its weighted adjacency matrix, we shall study the class of all real matrices of the same size as A and with entries having the same signs as the corresponding entries of A (and having 0's in the same places as A). This class will be denoted Q_A. The classes Q_A will be called *qualitative matrices*, to use the terminology of Maybee and Quirk (1969). We shall be interested in those properties which, if they hold for one matrix in a class Q_A, hold for all matrices in this class.

We have already mentioned a number of questions which can be formulated in this language. In particular, in Chapter 9 we talked about the eventual stability of sequences of matrices representing sentiments of members of a small group, and asked when eventual stability was a property of sign pattern alone. In Chapter 10, we asked when pulse or value stability was a property of sign pattern alone. Similar questions arise in economics (Quirk and Ruppert (1965), Samuelson (1947)), in ecology (Jeffries (1974), Levins (1974a,b), May (1973a,b)) and in chemistry (Clarke (1975), Tyson (1975)).

Corresponding to the class Q_A will be a signed digraph D_A. This signed digraph has vertices the rows of A, an arc from row i to row j if and only if the (i, j) entry of A is nonzero, and a sign $+$ on the arc (i, j) if and only if this entry is positive. We shall usually study the class Q_A by using the digraph D_A.

11.1. Sign solvability. A common problem encountered in social or biological contexts is to solve a system of linear equations

$$Ax = b, \tag{11.1}$$

where A is a square matrix and b is a vector. In many situations, the entries of the matrix A and of the vector b are known only *qualitatively*, i.e., up to sign. Suppose we know that the system (11.1) has a solution. Is the sign pattern of the solution determined from the sign patterns of A and b? If not, when is it so determined? Apparently Samuelson (1947) was the first to propose this question in an economic context. Using the notation defined above, we ask the following: if $Ax = b$, if $A' \in Q_A$, if $b' \in Q_b$, and if $A'x' = b'$, is x' necessarily in Q_x? If so, we say that the system (11.1) is *sign solvable*.

The problem of characterizing sign solvable systems was discussed at length by Lancaster (1962), (1964), (1965) and by Gorman (1964), and was solved by

Bassett, Maybee and Quirk (1968). Our discussion follows the latter and also Maybee and Quirk (1969).

We begin with two examples. Suppose

$$(11.2) \qquad A=\begin{pmatrix}-2 & 0\\ 1 & -5\end{pmatrix}, \qquad b=(10,10).$$

Then $Ax=b$ is sign solvable. For if $A'\in Q_A$ and $b'\in Q_b$, then

$$A'=\begin{pmatrix}-\alpha & 0\\ \beta & -\gamma\end{pmatrix}, \qquad b'=(\delta,\varepsilon),$$

for $\alpha, \beta, \gamma, \delta, \varepsilon$ positive. The equation $A'x'=b'$ can be written as

$$-\alpha x_1'=\delta, \qquad \beta x_1'-\gamma x_2'=\varepsilon.$$

Thus,

$$x_1'=-\frac{\delta}{\alpha}$$

and

$$x_2'=\frac{-(\beta\delta/\alpha)-\varepsilon}{\gamma}.$$

Both x_1' and x_2' are negative, independent of the values of α, β, γ, δ, and ε. We conclude that $Ax=b$ is sign solvable.

To give a second example, let

$$(11.3) \qquad A=\begin{pmatrix}-2 & 2\\ -2 & -2\end{pmatrix}, \qquad b=(2,2).$$

Then $x=\begin{pmatrix}-1\\ 0\end{pmatrix}$ is a solution to $Ax=b$. If $A'=A$ and $b'=(6,2)$, then $A'\in Q_A$, $b'\in Q_b$, and $x'=\begin{pmatrix}-2\\ 1\end{pmatrix}$ is a solution to $A'x'=b'$. However, $x'\notin Q_x$.

To state a characterization of sign solvability, let us first observe that if $\det A=0$, then one can show that the system is not sign solvable. Next, we observe that renumbering of equations or variables or multiplying equations and or variables by -1 does not affect sign solvability. Hence, if $\det A\neq 0$, one can show that the problem can be reduced to the case where $a_{ii}<0$, all i, $x_j\leqq 0$, all j, and $b_j\geqq 0$, all j. The characterization of sign solvability can now be stated in terms of the signed digraph D_A defined above.

THEOREM 11.1 (Bassett, Maybee and Quirk (1968)). *Suppose $a_{ii}<0$, all i, $x_j\leqq 0$, all j, and $b_j\geqq 0$, all j. Then the equation $Ax=b$ is sign solvable if and only if*

1) *every cycle of the signed digraph D_A is negative*

and

2) *whenever $b_j>0$, then every simple path from $k\neq j$ to j in D_A is positive.*

To illustrate Theorem 11.1, let us consider A and b of (11.3). The diagonal entries of A are negative, $b_j\geqq 0$, each j, and $x=\begin{pmatrix}-1\\ 0\end{pmatrix}$ is a solution with $x_j\leqq 0$, all j. Hence, Theorem 11.1 applies. The signed digraph D_A is shown in Fig. 11.1. Every cycle of D_A is negative. However, $b_1 > 0$ and 2, 1 is a negative path into vertex 1. Thus, condition 2) of the theorem is violated.

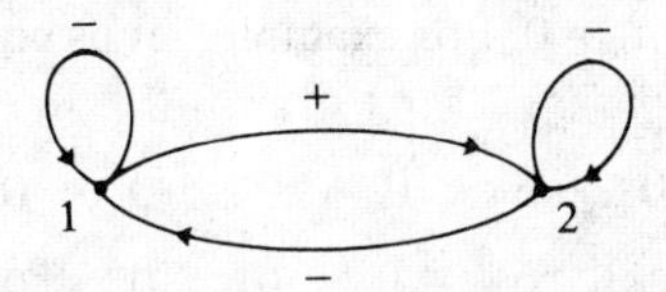

FIG. 11.1. *Signed digraph D_A corresponding to matrix A of* (11.3).

11.2. Sign stability. Another system of equations which arises with great frequency in economic and biological applications is the system of linear differential equations

$$\dot{x} = Ax. \tag{11.4}$$

We call an equilibrium $x = 0$ of the system (11.4) *stable* if for every $\varepsilon > 0$, there is a $\delta > 0$ with the property that every positive half-trajectory starting within δ of the origin lies eventually within ε of the origin. A well-known result states that the equilibrium $x = 0$ is stable if and only if every eigenvalue of A has negative real part. (See for example Cesari (1971) or Hahn (1967).) We call the matrix A *stable* if the equilibrium $x = 0$ of the system (11.4) is stable, i.e., if every eigenvalue of A has negative real part. We shall be interested in whether stability (in this sense, not in the sense of Chapter 10) can be inferred simply from the sign pattern, i.e., if stability is a property of the qualitative matrix. More generally, we shall be interested in a characterization of the *sign stable matrices*, the matrices A with the property that all matrices in Q_A are stable.

To give an example of a sign stable matrix, consider

$$A = \begin{pmatrix} -1 & 1 \\ -1 & -1 \end{pmatrix}. \tag{11.5}$$

If $A' \in Q_A$, then

$$A' = \begin{pmatrix} -\alpha & \beta \\ -\gamma & -\delta \end{pmatrix},$$

where α, β, γ, and δ are positive. The characteristic polynomial of A' is

$$\lambda^2 + \lambda(\alpha + \delta) + \alpha\delta + \beta\gamma = a\lambda^2 + b\lambda + c.$$

Using the quadratic formula, we see that if $b^2 - 4ac \leqq 0$, then each eigenvalue has real part $-(\alpha + \delta)/2$, a negative number. Similarly, since $ac > 0$, it is easy to verify that the real part is negative even if $b^2 - 4ac > 0$. Hence, each A' in Q_A is stable, and A is sign stable.

Quirk and Ruppert (1965) characterize sign stability if $a_{ii} \neq 0$, all i. A characterization in the general case was discovered by Jeffries (1974) and is discussed in Jeffries, Klee and van den Driessche (1977) and in Klee and van den Driessche (1977). We present Jeffries' result.[20]

To state this result, let us associate with the matrix A a graph G_A. The vertices of G_A are the rows of A, and there is an edge between rows i and j if and only if $i \neq j$ and both $a_{ij} \neq 0$ and $a_{ji} \neq 0$. For example, let us consider the matrix

$$A = \begin{pmatrix} 0 & -5 & 0 & 0 & 0 & 0 & 0 \\ 1 & 0 & -1 & 0 & 0 & 0 & 0 \\ 0 & 4 & -6 & -1 & 0 & -2 & 0 \\ 0 & 0 & 1 & 0 & -1 & 0 & 0 \\ 0 & 0 & 0 & 1 & 0 & 0 & 0 \\ 0 & 0 & 2 & 0 & 0 & 0 & -8 \\ 0 & 0 & 0 & 0 & 0 & 1 & 0 \end{pmatrix}. \tag{11.6}$$

The signed digraph D_A and graph G_A corresponding to A are shown in Fig. 11.2. Let

$$R_A = \{i : a_{ii} \neq 0\}.$$

In our example, the set R_A consists of the vertex 3 alone.

We shall try to color the vertices of G_A using two colors, white and black, in such a way that the following conditions are satisfied:

1) every vertex of R_A is black;
2) no black vertex has precisely one white neighbor;
3) every white vertex has at least one white neighbor.

Such a coloring, if it exists, is called an R_A-*coloring* of G_A. An R_A-coloring of G_A in our example is obtained by coloring vertex 3 black and all other vertices white, as shown in Fig. 11.2.

A *matching* of a graph is a set of pairwise disjoint edges of the graph. If S is a set of vertices in a graph G, an S-*complete matching* is a set M of pairwise disjoint edges of G such that all vertices not covered by the edges in M are outside S. For example, if $S = V - R_A$, then an S-complete matching in the graph G_A of Fig. 11.2 is given by the edges $\{1, 2\}$, $\{4, 5\}$ and $\{6, 7\}$.

THEOREM 11.2 (Jeffries). *An $n \times n$ real matrix A is sign stable if and only if the following conditions hold*:

1) *Each loop of the signed digraph D_A is negative.*
2) *Each cycle of length 2 in D_A is negative.*
3) *D_A has no cycles of lengths larger than 2.*
4) *In every R_A-coloring of the graph G_A, all vertices are black.*
5) *G_A has a $(V - R_A)$-complete matching.*

[20] Earlier results claimed in the general case by Quirk and Ruppert (1965) and Maybee and Quirk (1969) were shown to be incorrect in Jeffries, Klee and van den Driessche (1977).

The matrix A of (11.6) satisfies conditions 1), 2), 3), and 5), but violates condition 4), and hence we see that it is not sign stable. As a further illustration of this theorem, it is easy to check that if A is the matrix of (11.5), then all five conditions of Theorem 11.2 are satisfied.

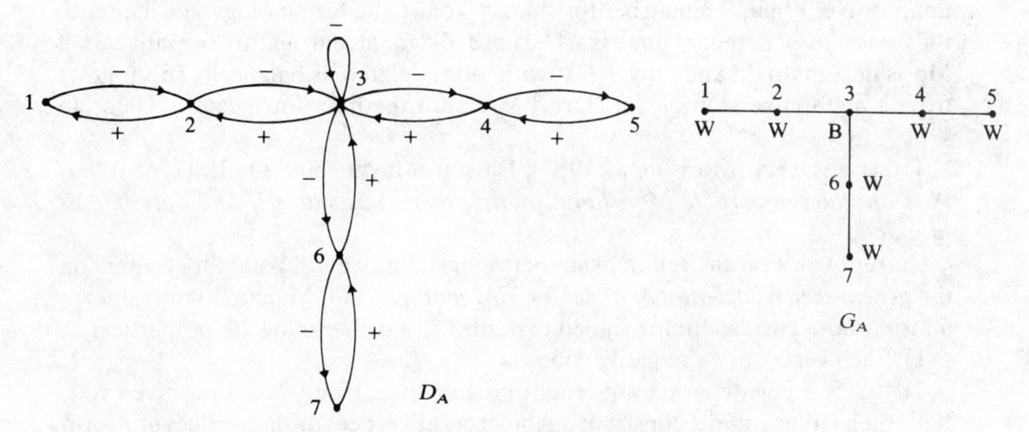

FIG. 11.2. *Signed digraph D_A and graph G_A corresponding to matrix A of* (11.6), *and an R_A-coloring of G_A.*

11.3. GM Matrices. A real $n \times n$ matrix A is called *perfectly stable*, or *Hicks stable*, or *H-stable*, if each principal minor A_p of order p satisfies $(-1)^p A_p > 0$. This notion was introduced in a famous book on economic theory by Hicks (1939). Since then, there has been a considerable body of literature devoted to the study of the relationship between H-stability and the notion of stability defined in the previous section. (This literature owes much of its impetus to a paper by Samuelson (1944) which argued that Hicks' notion of stability was not as relevant to economics as the notion of stability of § 11.2.) In particular, a number of authors have considered the problem of characterizing those matrices A for which the two notions of stability are equivalent. We shall discuss this problem, following the development of Maybee and Quirk (1973).

We shall discuss conditions on the signed digraph D_A which are sufficient to conclude that the two notions of stability are equivalent.

THEOREM 11.3 (Metzler (1945)). *Suppose that the signed digraph D_A satisfies the following conditions:*

1) *There is a negative loop at each vertex (i.e., $a_{ii} < 0$ for all i).*
2) *There are no other negative arcs (i.e., $a_{ij} \geqq 0$ for all $i \neq j$.)*
3) *D_A is strongly connected (A is "indecomposable").*

Then A is stable if and only if A is H-stable.

A matrix A such that D_A satisfies conditions 1) and 2) of Theorem 11.3 is called *Metzlerian*. The next class of matrices of interest generalizes the Metzlerian ones. We say that a real $n \times n$ matrix A is a *Morishima matrix* if for some permutation matrix P,

$$P^{\mathrm{T}}AP = \begin{pmatrix} A_{11} & A_{12} \\ A_{21} & A_{22} \end{pmatrix}, \tag{11.7}$$

where A_{11} and A_{22} are Metzlerian matrices and all entries of A_{12} and A_{21} are nonpositive. When is a matrix Morishima? To use the terminology of Chapter 9, it is easy to see that a matrix A whose diagonal entries are negative is a Morishima matrix if and only if D_A (with loops deleted) is balanced. This follows from a digraph version of the Harary structure theorem for balance (Theorem 9.1).

THEOREM 11.4 (Morishima (1952), Bassett, Maybee and Quirk (1968)).[21] *If A is an indecomposable Morishima matrix, then A is stable if and only if A is H-stable.*

Current work on the relationship between stability and H-stability centers on the *generalized Metzlerian matrices*, or *GM matrices*. A GM matrix is a real $n \times n$ matrix whose corresponding signed digraph D_A has the following properties:

1) Each vertex has a negative loop.
2) If I is a negative cycle of length greater than 1 and J is a positive cycle, then either I and J consist of disjoint sets of vertices or the vertices of I form a subset of the vertices of J.

It is easy to see that every Metzlerian matrix and every Morishima matrix is a GM matrix. The latter follows since D_A less loops is balanced and so has no negative cycles. It is easy to give examples of GM matrices which are not Morishima matrices. Maybee and Quirk (1973) show that for indecomposable GM matrices with all cycles negative, stability implies H-stability. Now if A is an indecomposable GM matrix with all cycles negative, then every $B \in Q_A$ has these properties. Hence, for indecomposable GM matrices A with all cycles negative, stability implies H-stability for all $B \in Q_A$. Maybee and Quirk also show that if A is an indecomposable matrix with negative elements on the principal diagonal, then, if A is not a GM matrix, it must be the case that for some $B \in Q_A$, B is stable but not H-stable. This suggests the conjecture, made by Maybee and Quirk, that among qualitatively specified real matrices A, stability implies H-stability if and only if A is a GM matrix. That is, stability implies H-stability for all $B \in Q_A$ if and only if A is a GM matrix. Our major reason for stating this conjecture here is that it suggests questions of graph-theoretic interest. Namely, what are the properties of the signed digraphs D_A corresponding to GM matrices? Such signed digraphs are called *GM digraphs*. Figure 11.3 shows an example of such a digraph. Verification of properties 1) and 2) is left to the reader. Maybee and Quirk (1973) obtain a variety of results about GM digraphs, of which a typical result says the following: if D is a strongly connected GM digraph with a positive and negative cycle going through the same vertex set, then every positive cycle goes through all the vertices. This result is illustrated by the GM digraph of Fig. 11.3. To the author's knowledge, no nice characterization of GM digraphs has been obtained. Such a

[21] Morishima's proof requires an additional hypothesis, namely that $a_{ij} \neq 0$, all i, j.

characterization might help in the solution of a long-standing problem in economics, namely the problem of determining the relationship between stability and H-stability.

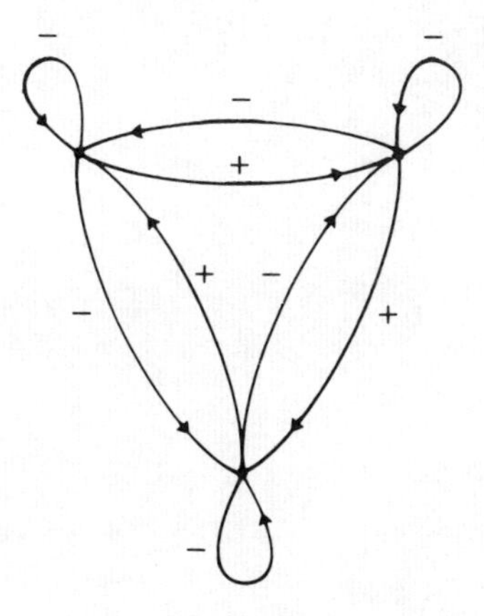

FIG. 11.3. *A GM digraph.*

References

L. G. ANTLE AND G. P. JOHNSON (1973), *Integration of policy simulation, decision analysis, and information systems: Implications of energy conservation and fuel substitution measures on inland waterway traffic*, Proceedings of Computer Science and Statistics: Seventh Annual Symposium on the Interface, Iowa State University, Ames.

K. APPEL AND W. HAKEN (1977), *Every planar map is four colorable. Part I: Discharging*, Illinois J. Math., 21, pp. 429–490.

K. APPEL, W. HAKEN AND J. KOCH (1977), *Every planar map is four colorable. Part II: Reducibility*, Ibid., 21, pp. 491–567.

W. E. ARMSTRONG (1939), *The determinateness of the utility function*, Econom. J., 49, pp. 453–467.

——— (1948), *Uncertainty and the utility function*, Ibid., 58, pp. 1–10.

——— (1950), *A note on the theory of consumer's behavior*, Oxford Economic Papers, 2, pp. 119–122.

——— (1951), *Utility and the theory of welfare*, Ibid., 3, pp. 259–271.

R. M. AXELROD, ED. (1976), *Structure of Decision*, Princeton University Press, Princeton, NJ.

L. BASSETT, J. MAYBEE AND J. QUIRK (1968), *Qualitative economics and the scope of the correspondence principle*, Econometrica, 36, pp. 544–563.

E. BELTRAMI AND L. BODIN (1973), *Networks and vehicle routing for municipal waste collection*, Networks, 4, pp. 65–94.

S. BENZER (1959), *On the topology of the genetic fine structure*, Proc. Nat. Acad. Sci. USA, 45, pp. 1607–1620.

——— (1962), *The fine structure of the gene*, Sci. Amer., 206, pp. 70–84.

C. BERGE (1961), *Färbung von Graphen, deren sämtliche bzw. deren ungerade Kreise starr sind*, Wiss. Z. Martin-Luther-Univ. Halle-Wittenberg, Math-Wiss. Reihe, 114.

——— (1962), *Sur une conjecture relative au problème des codes optimaux*, Comm. 13ème assemblee generale de l'URSI, Tokyo, International Scientific Radio Union.

——— (1963), *Perfect graphs*, I. Six Papers on Graph Theory, Indian Statistical Institute, Calcutta.

——— (1967), *Some classes of perfect graphs*, Graph Theory and Theoretical Physics, F. Harary, ed., Academic Press, New York.

——— (1969), *The rank of a family of sets and some applications to graph theory*, Recent Progress in Combinatorics, W. T. Tutte, ed., Academic Press, New York, pp. 49–57.

——— (1973), *Graphs and Hypergraphs*, American Elsevier, New York.

J. BERGER, M. H. FISEK, R. Z. NORMAN AND M. ZELDITCH (1977), *Status Characteristics and Social Interaction*, Elsevier, New York.

J. BERGER, M. ZELDITCH, B. ANDERSON AND B. P. COHEN (1972), *Structural aspects of distributive justice: A status value formulation*, Sociological Theories in Progress, II, J. Berger, M. Zelditch, and B. Anderson, eds., Houghton-Mifflin, Boston, pp. 119–146.

F. BOESCH AND R. TINDELL (1977), *Robbins' Theorem for mixed graphs*, unpublished manuscript, Bell Telephone Laboratories, Holmdel, New Jersey (submitted to Amer. Math. Monthly).

D. CARTWRIGHT AND F. HARARY (1956), *Structural balance: A generalization of Heider's theory*, Psych. Rev., 63, pp. 277–293.

D. B. CEARLOCK (1977), *Common properties and limitations of some structural modeling techniques*, Doctoral dissertation, Univ. of Washington, Seattle.

L. CESARI (1971), *Asymptotic Behavior and Stability Problems in Ordinary Differential Equations*, Springer-Verlag, New York-Heidelberg-Berlin.

W. K. CHEN (1967), *On directed graph solutions of linear algebraic equations*, SIAM Rev., 9, pp. 692–707.

——— (1971), *Applied Graph Theory*, American Elsevier, New York.

V. CHVÁTAL, M. R. GAREY AND D. S. JOHNSON (1976), *Two results concerning multicoloring*, Rep. STAN-CS-76-582, Computer Science Dept., Stanford Univ., Stanford, CA, December.

V. CHVÁTAL AND C. THOMASSEN (to appear), *Distances in orientations of graphs*, J. Combinatorial Theory Ser. B.

B. L. CLARKE (1975), *Theorems on chemical network stability*, J. Chem. Phys., 72, pp. 773–775.

F. H. CLARKE AND R. E. JAMISON (1976), *Multicolorings, measures and games on graphs*, Discrete Math., 4, pp. 241–246.

S. K. COADY, G. P. JOHNSON AND J. M. JOHNSON (1973), *Effectively conveying results: A key to the usefulness of technology assessment*, paper delivered at the First International Congress on Technology Assessment, The Hague, May 31, Institute for Water Resources, Corps of Engineers, Fort Belvoir, VA.

J. E. COHEN (1978), *Food Webs and Niche Space*, Princeton University Press, Princeton, N.J.

S. A. COOK (1971), *The complexity of theorem-proving procedures*, Proc. Third ACM Symposium on the Theory of Computing, Assoc. Comput. Mach., New York, pp. 151–158.

C. H. COOMBS AND J. E. K. SMITH (1973), *On the detection of structures in attitudes and developmental processes*, Psych. Rev., 80, pp. 337–351.

G. A. DIRAC (1961), *On rigid circuit graphs*, Abh. Math. Sem. Univ. Hamburg, 25, pp. 71–76.

J. EDMONDS (1965), *Minimum partition of a matroid into independent subsets*, J. Res. Nat. Bur. Standards Sect. B, B69, pp. 67–72.

D. R. FULKERSON AND O. A. GROSS (1965), *Incidence matrices and interval graphs*, Pacific J. Math., 15, pp. 835–855.

H. GABAI (1974), *N-dimensional interval graphs*, York College, City Univ. of New York, New York.

M. R. GAREY AND D. S. JOHNSON (1976), *The complexity of near-optimal graph coloring*, J. Assoc. Comput. Mach., 23, pp. 43–49.

M. R. GAREY, D. S. JOHNSON AND L. STOCKMEYER (1976), *Some simplified NP-complete problems*, Theoret. Comput. Sci., 1, pp. 237–267.

A. GHOUILA-HOURI (1962), *Caractérisation des graphes nonorientés dont on peut orienter les arêtes de manière a obtenir le graphe d'une relation d'ordre*, C.R. Acad. Sci. Paris Ser. A-B, 254, pp. 1370–1371.

E. N. GILBERT (1972), Unpublished technical memorandum, Bell Telephone Laboratories, Murray Hill, NJ.

P. C. GILMORE AND A. J. HOFFMAN (1964), *A characterization of comparability graphs and of interval graphs*, Canad. J. Math., 16, pp. 539–548.

M. C. GOLUMBIC (1977), *Comparability graphs and a new matroid*, J. Combinatorial Theory Ser. B, 22B, pp. 68–90.

I. J. GOOD (1947), *Normal recurring decimals*, J. London Math. Soc., 22, pp. 167–169.

T. GORMAN (1964), *More scope for qualitative economics*, Rev. Economic Studies, 31, pp. 65–68.

J. GRIGGS AND D. B. WEST (1977), *Extremal values of the interval number of a graph*, California Inst. of Tech., Pasadena, CA.

W. HAHN (1967), *Stability of Motion*, Springer-Verlag, Berlin-Heidelberg-New York.

A. HAJNAL AND J. SURÁNYI (1958), *Über die Auflösung von Graphen in Vollständige Teilgraphen*, Ann. Univ. Sci. Budapest. Eötvös Sect. Math., 1, pp. 113–121.

G. HAJOS (1957), *Über eine Art von Graphen*, Internat. Math. Nachr., 47, p. 65.

F. HARARY (1954), *On the notion of balance of a signed graph*, Michigan Math. J., 2, pp. 143–146.

——— (1959a), *A graph-theoretic method for the complete reduction of a matrix with a view toward finding its eigenvalues*, J. Math. and Physics, 38, pp. 104–111.

——— (1959b), *On the group of the composition of two graphs*, Duke Math. J., 26, pp. 29–34.

——— (1959c), *On the measurement of structural balance*, Behavioral Sci., 4, pp. 316–323.

——— (1969), *Graph Theory*, Addison-Wesley, Reading, MA.

F. HARARY, R. Z. NORMAN AND D. CARTWRIGHT, (1965), *Structural Models: An Introduction to the Theory of Directed Graphs*, John Wiley, New York.

J. L. HARRISON (1962), *The distribution of feeding habits among animals in a tropical rain forest*, J. Animal Ecology, 31, pp. 53–63.

F. HEIDER (1946), *Attitudes and cognitive organization*, J. Psychology, 21, pp. 107–112.

J. HICKS (1939), *Value and Capital*, Oxford University Press, Oxford.

F. S. HILLIER AND G. J. LIEBERMAN (1974), *Introduction to Operations Research*, Holden-Day, San Francisco, 1st ed. 1967.

A. J. W. HILTON, R. RADO AND S. H. SCOTT (1973), *A (<5)-colour theorem for planar graphs*, Bull. London Math. Soc., 5, pp. 302–306.

C. H. HUBBELL, E. C. JOHNSEN AND M. MARCUS (1978), *Structural balance in group networks*, Handbook of Social Science Methods, B. Anderson and R. B. Smith, eds., Irvington Publishers, distributed by Halsted Press, New York.

L. HUBERT (1974), *Some applications of graph theory and related non-metric techniques to problems of approximate seriation: The case of symmetric proximity measures*, British J. Math. Statist. Psychology, 27, pp. 133–153.

G. HUTCHINSON (1969), *Evaluation of polymer sequence fragment data using graph theory*, Bull Math. Biophys., 31, pp. 541–562.

G. E. HUTCHINSON (1944), *Limnological studies in Connecticut. VII. A critical examination of the supposed relationship between phytoplankton periodicity and chemical changes in lake waters*, Ecology, 25, pp. 3–26.

J. P. HUTCHINSON AND H. S. WILF (1975), *On eulerian circuits and words with prescribed adjacency patterns*, J. Combinatorial Theory Ser. A, A18, pp. 80–87.

C. JEFFRIES (1974), *Qualitative stability and digraphs in model ecosystems*, Ecology, 55, pp. 1415–1419.

C. JEFFRIES, V. KLEE AND P. VAN DEN DRIESSCHE (1977), *When is a matrix sign stable?* Canad. J. Math., 29, pp. 315–326.

J. KANE (1972), *A primer for a new cross-impact language—KSIM*, Tech. Forecasting and Social Change, 4, pp. 129–142.

——— (1973), *Intuition, policy and mathematical simulation*, paper presented at International Symposium on Uncertainties in Hydrologic and Water Resource Systems, University of British Columbia, Vancouver.

J. KANE, W. THOMPSON AND I. VERTINSKY (1972), *Health care delivery: A policy simulator*, Socio-Econ. Plan. Sci., 6, pp. 283–293.

J. KANE, I. VERTINSKY AND W. THOMPSON (1972), *Environmental simulation and policy formulation—methodology and example (water policy for British Columbia)*, International Symposium on Modeling Techniques in Water Resources Systems, A. K. Biswas, ed., Environment, Ottawa, Canada.

——— (1973), *KSIM: A methodology for interactive resource policy simulation*, Water Resources Research, 9, pp. 65–79.

R. M. KARP (1972), *Reducibility among combinatorial problems*, Complexity of Computer Computations, R. E. Miller and J. W. Thatcher, eds., Plenum Press, New York.

D. G. KENDALL (1963), *A statistical approach to Flinders Petrie's sequence dating*, Bull. Internat. Statist. Inst., 40, pp. 657–680.

——— (1969a), *Incidence matrices, interval graphs, and seriation in archaeology*, Pacific J. Math., 28, pp. 565–570.

——— (1969b), *Some problems and methods in statistical archaeology*, World Archaeology, 1, pp. 61–76.

——— (1971a), *A Mathematical approach to seriation*, Philos. Trans. Roy. Soc. London Ser. A, 269, pp. 125–135.

——— (1971b), *Abundance matrices and seriation in archaeology*, Z. Wahrscheinlichkeitstheorie und Verw. Gebiete, 17, pp. 104–112.

——— (1971c), *Seriation from abundance matrices*, Mathematics in the Archaeological and Historical Sciences, F. R. Hodson et al., eds., Edinburgh University Press, Edinburgh.

V. KLEE (1969), *What are the intersection graphs of arcs in a circle*?, Amer. Math. Monthly, 76, pp. 810–813.

V. KLEE AND P. VAN DEN DRIESSCHE (1977), *A linear-time test for the sign stability of a matrix*, Numer. Math., 28, pp. 273–285.

D. KÖNIG (1936), *Theorie des Endlichen und Unendlichen Graphen*, Akademische Verlagsgesellschaft M.B.H., Leipzig; reprinted, Chelsea, New York, 1950.

P. G. KRUZIC (1973a), *A suggested paradigm for policy planning*, Stanford Research Institute Tech. Note TN-OED-016, Menlo Park, CA, June.

——— (1973b), *Cross-impact analysis workshop*, Stanford Research Institute Letter Rep., Menlo Park, CA, June 23.

K. J. LANCASTER (1962), *The scope of qualitative economics*, Rev. Economic Studies, 29, pp. 99–123.

——— (1964), *Partionable systems and qualitative economics*, Ibid., 31, pp. 69–72.

——— (1965), *The theory of qualitative linear systems*, Econometrica, 33, pp. 395–408.

H. G. LANDAU (1955), *On dominance relations and the structure of animal societies: III. The condition for a score structure*, Bull. Math. Biophys., 15, pp. 143–148.

C. B. LEKKERKERKER AND J. CH. BOLAND (1962), *Representation of a finite graph by a set of intervals on the real line*, Fund. Math., 51, pp. 45–64.

G. G. LENDARIS AND W. W. WAKELAND (1977), *Structural modeling—a bird's eye view*, Systems Science Ph.D. Program, Portland State Univ., Portland, OR, February.

J. H. LEVINE (1976), *The network of corporate interlocks in the United States: An overview*, Dartmouth College, Hanover, NH.

R. LEVINS (1974a), *Problems of signed digraphs in ecological theory*, Ecosystem Analysis and Prediction, S. Levin, ed., Soc. Indust. Appl. Math., Philadelphia, pp. 264–277.

——— (1974b), *The qualitative analysis of partially specified systems*, Ann. New York Acad. Sci., 231, pp. 123–138.

T. M. LIEBLING (1970), *Graphentheorie in Planungs-und Tourenproblemen*, Lecture Notes in Operations Research and Mathematical Systems No. 21, Springer-Verlag, Berlin-Heidelberg-New York.

C. L. LIU (1968), *Introduction to Combinatorial Mathematics*, McGraw-Hill, New York.

——— (1972), *Topics in Combinatorial Mathematics*, Mathematical Association of America, Washington, D.C.

L. LOVÁSZ (1972a), *Normal hypergraphs and the perfect graph conjecture*, Discrete Math., 2, pp. 253–267.

——— (1972b), *A characterization of perfect graphs*, J. Combinatorial Theory, 13, pp. 95–98.

——— (1977), *On the Shannon capacity of a graph*, Preprint of the Mathematical Institute of the Hungarian Academy of Sciences, Budapest.

R. D. LUCE (1956), *Semiorders and a theory of utility discrimination*, Econometrica, 24, pp. 178–191.

E. MARCZEWSKI (1945), *Sur deux propriétés des classes d'ensembles*, Fund. Math., 33, pp. 303–307.

R. M. MAY (1973a), *Qualitative stability in model ecosystems*, Ecology, 54, pp. 638–641.

——— (1973b), *Stability and Complexity in Model Ecosystems*, Princeton University Press, Princeton, N.J.

J. MAYBEE AND J. QUIRK (1969), *Qualitative problems in matrix theory*, SIAM Rev., 11, pp. 30–51.

——— (1973), *The GM-matrix problem*, Tech. Rep., Dept. of Computer Sci., Univ. of Colorado, Boulder; Linear Algebra and Appl., to appear.

M. MCLEAN (1976), *Getting the problem right—the role for structural modeling*, Science Policy Research Unit, Univ. of Sussex, England.

M. MCLEAN AND P. SHEPHERD (1976), *The importance of model structure*, Futures, 8, pp. 40–51.

——— (1978a), *Feedback processes in dynamic models*, Tech. Forecasting and Social Change, 11, pp. 153–164.

——— (1978b), *The United Kingdom private car model—an example of structural modeling*, Science Policy Research Unit, Univ. of Sussex, England (submitted to Tech. Forecasting and Social Change).

K. MENGER (1951), *Probabilistic theories of relations*, Proc. Nat. Acad. Sci., 37, pp. 178–180.

L. METZLER (1945), *Stability of multiple markets: The Hicks conditions*, Econometrica, 13, pp. 277–292.

R. S. MILLER (1967), *Pattern and process in competition*, Advances in Ecological Research, Vol. 4, J. B. Cragg, ed., Academic Press, London, pp. 1–74.

M. MORISHIMA (1952), *On the laws of change of the price system in an economy which contains complementary commodities*, Osaka Economic Papers, 1, pp. 101–113.

J. MOSIMANN (1968), *Elementary Probability for the Biological Sciences*, Appleton-Century-Crofts, New York.

J. E. MOSIMANN, M. B. SHAPIRO, C. R. MERRIL, D. F. BRADLEY AND J. E. VINTON (1966), *Reconstruction of protein and nucleic acid sequences: IV. The algebra of free monoids and the fragmentation stratagem*, Bull. Math. Biophys., 28, pp. 235–260.

R. Z. NORMAN AND F. S. ROBERTS (1972a), *A derivation of a measure of relative balance for social structures and a characterization of extensive ratio systems*, J. Mathematical Psychology, 9, pp. 66–91.

——— (1972b), *A measure of relative balance for social structures*, Sociological Theories in Progress, II, J. Berger, M. Zelditch, and B. Anderson, eds., Houghton-Mifflin, New York, pp. 358–391.

T. R. PARSONS AND R. J. LEBRASSEUR (1970), *The availability of food to different trophic levels in the marine food chain*, Marine Food Chains, J. H. Steele, ed., University of California Press, Berkeley and Los Angeles, pp. 325–343.

E. R. PIANKA (1976), *Competition and niche theory*, Theoretical Ecology: Principles and Applications, R. M. May, ed., Blackwell Scientific, Oxford, England, pp. 114–141.

D. PIMENTEL, L. E. HURD, A. C. BELLOTTI, M. J. FORSTER, I. N. OKA, O. D. SHOLES AND R. J. WHITMAN (1973), *Food production and the energy crisis*, Science, 182, pp. 443–449.

D. PIMENTEL, W. R. LYNN, W. K. MACREYNOLDS, M. T. HEWES AND S. RUSH (1974), *Workshop on research methodologies for studies of energy, food, man and environment, phase I*, Tech. Rep., Cornell University Center for Environmental Quality Management, Ithaca, New York, June.

J. QUIRK AND R. RUPPERT (1965), *Qualitative economics and the stability of equilibrium*, Rev. Economic Studies, 32, pp. 311–326.

E. M. REINGOLD, J. NIEVERGELT AND N. DEO (1977), *Combinatorial Algorithms: Theory and Practice*, Prentice-Hall, Englewood Cliffs, NJ.

M. RICHARDSON (1946), *On weakly ordered systems*, Bull. Amer. Math. Soc., 52, pp. 113–116.

H. E. ROBBINS (1939), *A theorem on graphs, with an application to a problem of traffic control*, Amer. Math. Monthly, 46, pp. 281–283.

F. S. ROBERTS (1968), *Representations of indifference relations*, Ph.D. thesis, Department of Mathematics, Stanford Univ., Stanford, CA, January.

——— (1969a), *Indifference graphs*, Proof Techniques in Graph Theory, F. Harary, ed., Academic Press, New York, pp. 139–146.

——— (1969b), *On the boxicity and cubicity of a graph*, Recent Progress in Combinatorics, W. T. Tutte, ed., Academic Press, New York, pp. 301–310.

——— (1971a), *Homogeneous families of semiorders and the theory of probabilistic consistency*, J. Mathematical Psychology, 8, pp. 248–263.

——— (1971b), *On the compatibility between a graph and a simple order*, J. Combinatorial Theory, 11, pp. 28–38.

——— (1971c), *Signed digraphs and the growing demand for energy*, Environment and Planning, 3, pp. 395–410.

——— (1973), *Building and analyzing an energy demand signed digraph*, Ibid., 5, pp. 199–221.

——— (1974), *Structural characterizations of stability of signed digraphs under pulse processes*, Graphs and Combinatorics, R. Bari and F. Harary, eds., Lecture Notes #406, Springer-Verlag, Berlin-Heidelberg-New York, pp. 330–338.

——— (1975), *Weighted digraph models for the assessment of energy use and air pollution in transportation systems*, Environment and Planning, 7, pp. 703–724.

——— (1976a), *Discrete Mathematical Models, with Applications to Social, Biological, and Environmental Problems*, Prentice-Hall, Englewood Cliffs, NJ.

——— (1976b), *Structural analysis of energy systems*, Energy: Mathematics and Models, F. S. Roberts, ed., Soc. Indust. Appl. Math., Philadelphia, pp. 84–101.

——— (1978), *Food webs, competition graphs, and the boxicity of ecological phase space*, Theory and Applications of Graphs—in America's Bicentennial Year, Y. Alavi and D. Lick, eds., Springer-Verlag, New York.

——— (to appear), *On the mobile radio frequency assignment problem and the traffic light phasing problem*, Proc. Second International Conference on Combinatorial Mathematics, New York Academy of Sciences, New York, 1978.

F. S. ROBERTS AND T. A. BROWN (1975), *Signed digraphs and the energy crisis*, Amer. Math. Monthly, 82, pp. 577–594.

W. S. ROBINSON (1951), *A method for chronologically ordering archaeological deposits*, American Antiquity, 16, pp. 293–301.

M. ROSENFELD (1967), *On a problem of C. E. Shannon in graph theory*, Proc. Amer. Math. Soc., 18, pp. 315–319.

——— (1970), *Graphs with a large capacity*, Proc. Amer. Math. Soc., 26, pp. 57–59.

P. SAMUELSON (1944), *The relation between Hicksian stability and true dynamic stability*, Econometrica, 12, pp. 256–257.

——— (1947), *Foundations of Economic Analysis*, Harvard University Press, Cambridge, MA, 2nd ed., 1955.

S. H. SCOTT (1975), *Multiple node colorings of finite graphs*, Ph.D. dissertation, Department of Mathematics, Univ. of Reading, England, March.

C. E. SHANNON (1956), *The zero-error capacity of a noisy channel*, IRE Trans. Information Theory, IT-2, pp. 8–19.

L. N. SHEVRIN AND N. D. FILIPPOV (1970), *Partially ordered sets and their comparability graphs*, Siberian Math. J., 11, pp. 497–509.

S. STAHL (1976), *n-tuple colorings and associated graphs*, J. Combinatorial Theory, 20, pp. 185–203.

L. STOCKMEYER (1973), *Planar 3-colorability is polynomial complete*, SIGACT News, 5, pp. 19–25.

K. E. STOFFERS, (1968), *Scheduling of traffic lights—A new approach*, Transportation Research, 2, pp. 199–234.

H. F. TAYLOR (1970), *Balance in Small Groups*, Van Nostrand Reinhold, New York.

W. T. TROTTER AND F. HARARY (1977), *On double and multiple interval graphs*, Univ. of South Carolina, Columbia (submitted to J. Graph Theory).

W. T. TROTTER, J. I. MOORE AND D. P. SUMNER (1976), *The dimension of a comparability graph*, Proc. Amer. Math. Soc., 60, pp. 35–38.

A. C. TUCKER (1970), *Characterizing circular-arc graphs*, Bull. Amer. Math. Soc., 75, pp. 1257–1260.

——— (1971), *Matrix characterizations of circular-arc graphs*, Pacific J. Math., 39, pp. 535–545.

——— (1973), *Perfect graphs and an application to optimizing municipal services*, SIAM Rev., 15, pp. 585–590.

A. C. TUCKER AND L. BODIN (1976), *A model for municipal street-sweeping operations*, Case Studies in Applied Mathematics, Committee on the Undergraduate Program in Mathematics, Mathematical Association of America, Washington, D.C.

J. TYSON (1975), *Classification of instabilities in chemical reaction systems*, J. Chem. Phys., 62, pp. 1010–1015.

J. H. VANDERMEER (1972), *Niche theory*, Annual Review of Ecology and Systematics, vol. 3, R. F. Johnston, ed., Annual Reviews, Palo Alto, CA, pp. 107–132.

J. VON NEUMANN AND O. MORGENSTERN (1944), *Theory of Games and Economic Behavior*, Princeton University Press, Princeton, NJ, 2nd ed. 1947, 3rd ed. 1953.

H. M. WAGNER (1975), *Principles of Operations Research with Applications to Managerial Decisions*, Prentice-Hall, Englewood Cliffs, NJ, 1st ed., 1969.

E. M. WILKINSON (1971), *Archaeological seriation and the traveling salesman problem*, Mathematics in the Archaeological and Historical Sciences, F. R. Hodson, et al., eds., Edinburgh University Press, Edinburgh.

M. ZELDITCH, J. BERGER AND B. P. COHEN (1966), *Stability of organizational status structures*, Sociological Theories in Progress, I, J. Berger, M. Zelditch, and B. Anderson, eds., Houghton-Mifflin, Boston, pp. 269–294.

Subject Index

Author Index